Hawaiian
Reef
Animals

Hawaiian
Reef
Animals

Edmund S. Hobson and E. H. Chave

1972 The University Press of Hawaii

Contents

Illustrations

Preface

This book was written to fulfill a long-standing need for a compendium of Hawaiian reef animals that meets the needs and interests of people living and visiting in Hawaii. Its preparation has been supported in part by the State of Hawaii under Bill 902, Act 106, and the book constitutes a part of the marine atlas authorized by the Act. Additional support was provided by the National Sea Grant Program NOAA, Department of Commerce.

All photographs were taken by the senior author, Ted Hobson, who also wrote the Introduction and the section on fishes. Deetsie Chave wrote the section on invertebrates.

Both authors researched and checked the original references. Numbers in parentheses in the text refer to references as they are numbered in the Bibliography. Hawaiian word usage and phonetics have been verified using the *Hawaiian Dictionary* by M. K. Pukui and S. H. Elbert, University of Hawaii Press, 1971. An especially rich source of information was *Native Use of Fish in Hawaii,* by Margaret Titcomb, with the collaboration of Mary Kawena Pukui.

We also acknowledge with thanks the assistance of several faculty members of the University of Hawaii, who provided information and counsel in their respective specialties: Mr. David Hurd, Mrs. Ruby Johnson, Drs. E. Alison Kay, John A. Maciolek, Albert L. Tester, and Sidney J. Townsley. Dr. John E. Randall, of the Bernice P. Bishop Museum, and the late John Kuahiwinui, of Kona, also contributed information.

Introduction

Hawaii, her land and the traditions of her people, has evolved intimately tied to the sea. The islands themselves erupted from the ocean floor in times so recent that signs of their birth remain everywhere, and a marine tradition permeates all of man's history on these bits of land. Until very recent times, travel from one place to another, even on a single island, was usually over water, and most settlements were, and still are, along the shores. Sea life has traditionally been the major source of protein for Hawaiians, and probably every edible marine organism accessible to their hands has been at one time or another a regular part of the diet. All forms of marine life—the fishes, corals, crustaceans, mollusks, even seaweeds— were grouped together by ancient Hawaiians as *i'a* (1).

All of the early Islanders were familiar with the reef's inhabitants, but it was with a select group of professional fishermen, the *po'o lawai'a* (2), that fishing developed as a fine art. These respected men, numbering both commoners and royalty in their ranks, took great pride in their thorough command of the fishing trade. The profession carried with it a vast legacy of knowledge. Each generation of fishermen augmented this legacy with knowledge gleaned from their experiences before passing it on to their own chosen apprentices. The legacy included not only knowledge of techniques and equipment, but also of the habits of the different marine creatures—their specific habitats, food preferences, spawning, and migrations. For the *po'o lawai'a*, fishing was not a haphazard game of chance, but a science in which the technique used to catch a certain species was derived from an understanding of the habits of that particular animal. As was custom, the spoken word and personalized instruction alone were used in the transmission of knowledge from one generation to the next. With the passing of the old ways, too many years passed before chroniclers of Hawaiian history, pen in hand, probed the fading memories of those few aged fishermen who remembered how it was. Consequently, there remains today only a fragmented picture of the fishing knowledge of the early Hawaiians.

There was of course no break in the use of fish as food. But as people from other cultures flooded into the Islands during the last half of the nineteenth century, the ways of fishing, like so many other Hawaiian customs, experienced a rapid change. The Japanese, especially, possessing a rich and proud tradition as fishermen in their own right, introduced many new techniques, and modified Hawaiian ones. For example, the throw net, which today approaches a cliché in its association with native Hawaiian fishermen, was actually unknown in the Islands until introduced by the Japanese during the 1890s (3). Thus the fishing practices of Hawaii today are a rich blend of the more durable practices of the early Islanders and of introduced techniques, many of which were modified to suit the Islands' unique conditions.

Today the role of fishes in early Hawaiian beliefs and social custom is better known than is the ancient fishing knowledge. Although generally diluted or discredited by exposure to the outer world, the folklore persists. Even when credence has been lost, many of the old beliefs have lived on if only because their telling makes a good story. To many of today's Islanders, the old beliefs are no more than quaint relics of another time; but among others, especially those of Hawaiian blood, many of the ancient beliefs are still deeply respected.

The ancient Hawaiians acknowledged many gods who, they believed, profoundly influenced even the most ordinary of their everyday activities. The god of fishes and fishermen was Kū'ulakai, who was worshiped along with his wife Hinapukui'a, and son 'Ai'ai (2). Small fishing shrines, called ko'a and usually dedicated to Kū'ulakai, were located near the water's edge. Many of these were sacred to certain fish. For example, on the 'Ewa side of Koko Head a shrine called Palialea was sacred to mullet (4). These were simple structures, frequently no more than a flat stone on a rocky point. In fact, so simple were these structures that the remaining ones easily go unnoticed by the uninitiated. An individual designated as the guardian of the ko'a offered here the first fish of each catch. Significantly, even though the heiau, temples of the old religion, have been deserted for over a century, some Hawaiians still place offerings at a ko'a, especially when fishing is poor.

Fishermen believed that Kū'ulakai controlled all fishes in the sea, and that the whim of this god determined fishing success. Beyond this, each fisherman also venerated a personal god, or 'aumakua. This deity inhabited the body of some individual plant or animal, and looked over, and protected, its devotee; fishermen usually found their 'aumākua in sea creatures. Certain individual sharks, in particular, were considered the physical embodiment of an 'aumakua. Eels, porpoises, and 'opihi were also among forms venerated by various fishermen (5). And whenever an abnormally colored or shaped fish was inadvertently taken, it was returned to the sea lest it prove to be an 'aumakua (2). Individual creatures recognized as 'aumākua were offered food and prayer daily at certain points along the shore; neglect of this duty invited disastrous consequences.

Certain fishes, considered among the most highly esteemed foods, were regarded as appropriate offerings to the gods; red or white fishes were most often used. Pigs were one of the most important ceremonial offerings, but frequently could not be obtained when needed. As Hawaiians believed that living things on land had their counterparts in the sea, certain fishes were recognized as equivalent to the pig and were used in its place. Such fishes were known as *pua'a kai,* or sea pigs. Further, Hawaiians believed magic to be inherent in the meanings of words, and, as many Hawaiian words have multiple meanings, frequently a fish was chosen as an offering because its name carried a meaning pertinent to the situation (2).

The ancient Hawaiians recognized that fishes are a resource that must be actively conserved. The fishermen knew there was a limit to the numbers of fish they could take from a given spot before it was no longer productive. Sadly, this concept has had little impact on the fishing practices of latter-day Islanders, and many species that remained abundant during generations of intensive, but controlled, fishing by early Islanders have become rare during the past century.

A major instrument of conservation in ancient Hawaii was the *kapu,* or forbidden practices (2). Among the many *kapu* that regulated the consumption of fishes was one that forbade the taking of many species during certain times of the year, periods that often corresponded to the spawning seasons of the species. The penalty for violating the *kapu* was severe; frequently, instant death awaited the individual who consumed a forbidden fish. And even should the

violator of a *kapu* escape detection by those enforcing the *kapu,* he knew that his actions were witnessed by the gods, and that surely they would deal with him in their own way. Not surprisingly, the fish and game laws of that time were far more closely adhered to than those of our day.

The old ways passed as newcomers from other lands and diverse cultures arrived to become Islanders themselves. The resulting hybrid culture was less attuned to the natural conditions of the Islands than had been the culture it supplanted. To this day, there have been earnest and continuing efforts to get back into step with nature, but success has been elusive. The accelerating complexity of present Hawaiian society has made the enormousness of this task staggering.

Today Hawaii is a marine science center. Internationally recognized authorities in a wide range of specialties, particularly those concerned with tropical seas, study in the Islands' various scientific establishments: the University of Hawaii, Bishop Museum, National Marine Fisheries Service Laboratory, Oceanic Institute, and the Fish and Game Division of the state's Department of Land and Natural Resources. Much of this study concerns marine life on Hawaiian reefs; indeed, not since the days of the *po'o lawai'a* has so much general knowledge of Hawaiian marine life been available.

In this book we strive to discuss those aspects of Hawaiian reef animals that have drawn the attention of Islanders past and present. We center on the role of these animals in early Hawaiian folklore and traditions, but also include topics that have excited the interest of more recent residents.

The Fishes

Among creatures of the sea, sharks have been especially meaningful to Hawaiian Islanders. In the Hawaiian language, sharks, in general, are called *manō*. More than any other marine animal, sharks were worshipped in ancient times as *'aumākua*. Generally, shark *'aumākua* were believed to have had human origins—many being traced to aborted human foetuses that had been cast into the sea (5). Perhaps owing to its appearance, the partly formed foetus may have suggested a supernatural union. The foetus was believed to return to its family as an *'aumakua*, spiritually embodied in the form of a shark. Thereafter, members of this family, including descendants, worshipped what they believed to be the individual shark that represented their *'aumakua*. To them, shark meat, as food, was *kapu*, and they were required to make regular offerings of food and prayer at the place along the shore where their shark *'aumakua* lived. In return, they believed, the shark would protect them from harm while at sea and also would bring them good fishing. Sharks were the *'aumākua* of families among both the royalty and commoners: the family of King Kalakaua found ther *'aumakua* in the form of a shark. Although regarded as a god, an *'aumakua* nevertheless was also considered a servant of the family.

Sometimes all the inhabitants of a coastal region regarded a single shark as their *'aumakua*, and its name, history, appearance, place of abode, and other individual characteristics were well known to all. A member of a certain family charged with performing the appropriate offerings was known as a *kahu*. The position of *kahu* was hereditary and handed down within a family from one generation to the next (5). The people believed that the *kahu* was endowed with special powers because of his intimate relation with the *'aumakua* and therefore he exerted great influence over the community. Although the shark *'aumakua* was generally thought to protect its devotees, the *kahu* was frequently feared by the community (5). Taking advantage of the power he held in the minds of the people, a *kahu* sometimes terrorized his neighbors by adorning himself with trappings that made him resemble a shark, and by speaking in a squeaky falsetto tone of voice (1). It was believed, for example, that a *kahu* could transmit disease to those who displeased him, and in a village struck by sickness the malady was often attributed to the local *kahu*.

It was believed that most shark *'aumākua* could take human form at will—hence sharks believed to be *'aumākua* were collectively called *manō kānaka*, or shark men. There are many old Hawaiian stories of shark men who, in their human form, looked like other men except in having the mouth of a shark on their backs, and who became sharks upon leaping into the sea. A typical story tells of Nenewe, a shark man, who lived beside a path to the sea in Waipi'o Valley, on the Big Island of Hawaii (6). Nenewe

Sharks

hailed swimmers and fishermen taking the path and inquired of their destination. Then, armed with this knowledge, he raced ahead on an alternate trail, and, taking his shark form in the sea, devoured his victims when they arrived. Because of such folklore, many Hawaiians even today are reluctant to reveal an intent to go fishing.

The submarine homes of the shark deities were revered by the early Islanders, and even into modern times these locations have been considered by many to be sacred places. The twentieth century assumed a collision course with ancient Hawaiian beliefs when, in 1913, the U.S. Navy began building the large drydock in Pearl Harbor. The site chosen was known to belong to the shark goddess Ka'ahupāhau, but the Navy ignored the warnings of old Hawaiians. Construction proceeded without incident until the drydock was nearly completed. Then, in a sudden roar of splintering timbers, the entire structure collapsed. Why? The official record is not clear. Nevertheless, when work was begun again a priest of the old religion, a *kahuna*, was enlisted to appease Ka'ahupāhau. After appropriate prayers, and a ceremonious offering, work on the drydock was resumed. This time the job was successfully completed. Later, when the water was pumped from the drydock, the remains of a 14-foot shark were found at the bottom (7). Those unwilling to accept the influence of Ka'ahupāhau in this incident can readily cite coincidence—but there are still many who sincerely believe otherwise.

The most common sharks on Hawaiian reefs are the various grey sharks of the family Carcharhinidae. Many old Hawaiian names probably refer to these sharks, but whether these names correspond to species recognized today, or just to certain individuals, is uncertain. Examples include *manō pā'ele* (black-smudged *manō*), *manō lele wa'a* (canoe-jumping *manō*), and *manō pahāha* (thick-necked *manō*) (2). An attempt to associate these names with the species recognized today would be purely conjectural.

It has long been recognized that several varieties of grey sharks live in Hawaiian waters. However, not until Dr. Albert L. Tester, senior professor at the University of Hawaii, began an intensive shark-fishing program around the major Islands during the last decade have enough specimens been placed into the hands of specialists to relate these sharks to those in other seas. Several earlier ichthyologists had assigned various names to Hawaiian grey sharks, generally regarding them to be found only in these Islands. As a result of Dr. Tester's program, it became recognized that all the species of grey sharks in Hawaii are wide-ranging in tropical seas.

The three species of grey sharks most often seen on Hawaiian reefs are the sandbar shark, *Carcharhinus milberti*; the blacktip shark, *C. limbatus*; and the Galapagos shark, *C. galapagensis*.

Of these, the sandbar shark is by far the most common. This shark, which is also a common species in the tropical western Atlantic, rarely exceeds 6 feet in Hawaii. An underwater swimmer will most often see the sandbar shark swimming close to the sea floor, and will find it to be a timid animal that scurries away upon meeting a human. The blacktip shark, which is widely distributed in tropical seas of the world, is usually somewhat larger, with those seen often being about 7 or 8 feet long. Mostly it swims several feet above the bottom at the outer edge of the reef. It is not a timid shark, yet upon meeting a human in the water it usually remains some distance away, and soon moves out of sight. The Galapagos shark was once thought limited to islands in the eastern Pacific, off the Americas, but is now known to occur around oceanic islands throughout most of the world's tropical seas. Although this shark may grow to over 12 feet long, most individuals seen on the reefs are much smaller than this. The shark that swims up to investigate a swimmer at the surface, or in midwater, often turns out to be a Galapagos shark. Such encounters simply display its curious nature, however, and upon satisfying this curiosity the Galapagos shark usually fades off into the blue water from which it appeared.

Other grey sharks, less abundant in waters around the major Hawaiian Islands than the three described above, nevertheless are seen frequently in some areas. These include the blacktipped reef shark, *Carcharhinus melanopterus*, and the grey reef shark, *C. menisorrah* (Plate 1). These two, neither of which ordinarily grows to over 6 or 7 feet long, are very numerous in the lagoons of coral atolls throughout Oceania, but seem less at home in Hawaii. Another species, the whitetipped reef shark, *Triaenodon obesus* (Plate 2), is considered here among the grey sharks, even though many ichthyologists classify this distinctive animal among another group of sharks. The whitetipped reef shark is also common on the reefs of Pacific coral atolls, but is not abundant in Hawaii. Nevertheless, because the species frequents shallow water, and tends to remain in localized areas, it is regularly seen on those relatively few Hawaiian reefs that it inhabits. For example, it is a familiar sight on the reefs off Ke'ei and Hōnaunau in Kona. Unlike the other grey sharks, which must swim constantly to keep oxygen-supplying water flowing over their gills, the whitetipped reef shark can pump water over its gills while it rests motionless on the bottom. A skin diver peering into a cave may be startled to discover one of these sharks resting in the shadows.

Two species of hammerheads occur in Hawaiian waters, the scalloped hammerhead, *Sphyrna lewini*, and the smooth hammerhead, *Sphyrna zygaena*. Both grow to over 12 feet long, but only the former is abundant. So far as is known, the early Hawaiians only recognized one hammerhead, referring to it as *manō kihikihi*. Al-

Plate 1. Grey reef shark
Carcharhinus menisorrah.

Plate 2. Whitetipped reef shark, *Triaenodon obesus.*

though usually frequenting deep water offshore, female scalloped hammerheads come into shallow, sheltered waters during the summer to have their young. At this time they are especially numerous in Kāne'ohe Bay, on Oahu. Otherwise, hammerheads are not ordinarily seen on the reefs.

Early Hawaiian stories often tell of *niuhi*, a species of man-eating shark that was much feared by the Islanders. There are references to the eyes of this shark being luminous at night, as in the old chant:

Niuhi with fiery eyes
That flamed in the deep blue sea.
Alas! and alas!
When flowers the wili-wili tree
That is the time when the shark-god bites.
Alas! I am seized by the huge shark!
O blue sea, O dark sea,
Foam-mottled sea of Kane!
What pleasure I took in my dancing!
Alas, now consumed by the monster shark. (8)

Fishing for *niuhi* was a great event in ancient times—the sport of royalty. In anticipation, a great store of bait was accumulated. Human flesh was a favorite bait with many chiefs, not only because it was easier to obtain than pig, but its use also provided the chiefs an opportunity to eliminate persons who had fallen from favor. Kamehameha I, a renowned shark hunter, kept his victims imprisoned near the great *heiau* of Mo'okini, near Kawaihae on the island of Ha-

waii. After the victims had been cut up, their flesh was allowed to decompose for several days, thus enhancing its effectiveness as shark bait (3).

One report describes details of an expedition to capture a *niuhi* (9). A fleet of canoes, laden with bait, and with the *po'o lawai'a* and a *kahuna* in the lead, sailed to waters known to be frequented by this great shark. Once at anchor on the fishing ground, they cast great quantities of bait into the sea, for only after the waters for miles around carried the scent—a process that usually took several days—could the *niuhi* be expected to appear. When it arrived, the monster was fed huge quantities of food, which had been mixed liberally with pounded *'awa*, a narcotic herb of the pepper family. Gradually the *niuhi* became lethargic to the point that a noose could be slipped around its head. This was the climax. Once this had been accomplished, the fleet raised anchor and headed home. Groggy with *'awa*, satiated with food, and tethered with a line, the now docile *niuhi* followed the canoes—periodically receiving *'awa*-laced food to ensure its continued cooperation. At home, the fishermen led the stupefied shark into shallow water, where it was stranded and finally killed. Its flesh was not eaten, but portions of its skeleton and skin were coveted by the fishermen, who believed that the possession of these objects would endow them with courage. And he who had actually placed the noose around the head of the *niuhi* was confident of being forever victorious in battle. Too, the teeth of this shark were prized as weapons.

In the literature of recent times it has been assumed that *niuhi* was the white shark, *Carcharodon carcharias*, and indeed it may well have been. But a good case can be made that it was instead the tiger shark, *Galeocerdo cuvieri*. White sharks are seen only rarely in Hawaiian waters today; unless they were more numerous in ancient times, it would not seem that they could have accounted for the attention generated by *niuhi*. Tiger sharks, on the other hand, are regular inhabitants of Island waters, and individuals over 14 feet long weighing a ton or more are not infrequently seen. Such a monster could not have failed to deeply impress early Islanders, and yet no Hawaiian name has been linked with the tiger shark, even though this distinctive animal has many specific morphological characteristics that set it apart from other sharks. Certainly the tiger shark is world renowned as a man-eater; indeed, of the sharks that regularly frequent Hawaiian waters, the tiger shark represents the greatest threat to humans.

Despite the strong shark tradition of the Islands, Hawaiian waters are among the safest in tropical seas. In the 85 years since records were first kept, only sixteen people have been injured by sharks—five fatally. This is remarkably few, considering the countless numbers of people who regularly have entered Island waters for work and play over the years.

During most of this time there was a viable shark fishery in Hawaii, with the catch being marketed largely as an ingredient of fish cakes. However, the fishery suffered a mortal blow in the early 1940s, when legislation was passed requiring that the ingredients of fish cakes be listed on the package, and consumers rejected the use of shark meat. The resulting demise of the fishery was both unnecessary and unfortunate: unnecessary because shark meat is in fact highly palatable fare when properly prepared, and unfortunate because with little doubt the fishery kept the numbers of sharks in Hawaiian coastal waters at a low level. Shark populations are especially vulnerable to fishing, compared to those of most other fishes. This is largely because the developing young are carried by the females for a long period, and because only a relatively few young are produced by each female during a breeding season. In the years following the end of the shark fishery, sightings of these animals increased, and it was the opinion of many experts that the numbers of sharks in Hawaiian coastal waters was on the upswing. Then in 1958 a tiger shark attacked and fatally injured a fifteen-year-old surfer on Windward Oahu. Public alarm crystalized with this incident, and support was generated leading to the several shark-fishing programs supervised by Dr. Tester over the next decade. In addition to much-needed information on shark biology, these efforts demonstrated that even a limited fishing effort, when systematically conducted, can effectively reduce the numbers of sharks that are active in nearshore waters.

Eels

Most eels in Hawaii are morays, of the family Muraenidae. Collectively called *pūhi* by Islanders, these predators are denizens of cracks and crevices in the reef. Because they venture only infrequently into the open, their great abundance cannot be appreciated by the casual observer. Those seen most often are the various species whose members characteristically protrude their heads from crevices in the reef. Others frequently encountered are those that are quick to detect injured fishes, and thus appear on the scene shortly after the spear of a skin diver has found its mark. But these sightings are less frequent than one might expect, considering that, after the wrasse family Labridae, the family Muraenidae has the largest number of fish species on Hawaiian reefs (10).

Moray eels were the *'aumākua* of many early Hawaiians (2). One well-known eel-god, called Ko'ona, was worshiped by the people of Wailau on the windward coast of Molokai. Many heroic deeds are attributed to this deity, and to this day a large cave on the rock shore is said to have been formed when Ko'ona caused the cliffside to fall on a large shark that had invaded the area. According to legend (11), Ko'ona met his end after raiding the fishponds of Kū'ulakai, god of fishermen, at Hāna, Maui. It was 'Ai'ai, son of Kū'ulakai, who led several canoes on the expedition that destroyed Ko'ona. A large hook was secured to a long line and baited. Weighting himself with stones, 'Ai'ai dived with the hook to the opening of a submarine cave known to harbor Ko'ona. When the huge *pūhi* took the bait, and was hooked, the canoes trailed the line to shore at Lehaula. Here it was taken up by the people of the area, who pulled together and hauled the giant *pūhi* onto the beach. Three *'a'ā* stones, hurled by 'Ai'ai at the stranded eel-god, provided the death blows. On the beach at this place today a rock formation 30 feet long is said to be the remains of Ko'ona's backbone, and another group of rocks awash in the sea a short distance away is claimed to have been the creature's jaw bones.

Despite the probably exaggerated reports by skin divers of giants over 10 feet long, *pūhi* on Hawaiian reefs today do not appear to exceed a length of more than about 5 or 6 feet. The larger eels are a small minority, most Hawaiian

10

species being not longer than about 2 feet when fully grown (10). An example is *Gymnothorax meleagris* (Plate 3), known by early Hawaiians as *pūhi ʻōniʻo* (spotted *pūhi*); this *pūhi* characteristically protrudes its head from the coral and is perhaps the moray seen most often by snorkelers swimming over the reef.

There are many Hawaiian names for the different *pūhi*, but, as is true of the grey sharks, these names are often difficult to match with the species recognized today. Such names were applied only to the larger members of a species; small individuals of all species were referred to collectively by several names: *pūhi ʻauʻaukī, pūhi puakī* (after *puakiʻi*, thin), or *pūhi ʻoīlo* (young *pūhi*).

The *pūhi* were heavily fished by early Hawaiians. Very small individuals were frequently caught by hand, a technique known as *ʻini ʻiniki pūhi* (2). In this method, the fisherman placed a small octopus, *heʻe pali*, in the palm of his left hand, allowing the tentacles to dangle between his fingers in rocky shallows where small *pūhi* were known to abound. When an eel emerged to seize one of the tentacles, the fisherman used his right hand to pull the octopus slowly from his palm and toward his wrist, an action that withdrew the tentacles from between his fingers. The small eel followed, and when the fisherman saw the eel's head appear between his fingers he snapped his fist closed, securing the tiny creature in his grasp. Often several eels

were captured simultaneously in this way, but understandably the method was not used to catch larger morays. In fact, when a big *pūhi* appeared fishermen often would leave the area.

The larger eels were fished, but the fishermen took pains to avoid injury. Hook and line, or a spear, were used. In wading depths, larger *pūhi* often were chased into scoop nets, whereupon the fisherman raced across the shallows, trying to reach shore before his writhing captive could struggle free (3). Once landed, *pūhi* were dispatched with two blows of a mallet—one blow to the head, the other to the tail (12). Because an eel frequently is stunned more effectively by a blow to the tail than to the head, many Island fishermen even now contend that an eel's brain is in its tail. This anomaly can be traced to the fact that the moray's brain is encased in an especially heavy and strong skull, an adaptation to the animal's habit of wedging its way among the narrow openings within the reef.

The larger eels were feared by Islanders, and justifiably so, because these animals can inflict serious injury. The most feared, *Gymnothorax flavimarginatus* (Plate 4), was called *pūhi paka* (fierce *pūhi*). This heavy-bodied species is the most frequently seen of the larger *pūhi* and was sought often by the more adventuresome fishermen. There are other large eels with more vicious dispositions, which are better equipped to inflict injury, for example the dark brown species *Enchelynassa canina*. Nevertheless, the *pūhi*

11

Plate 3. Moray eel, *Gymnothorax meleagris* (pūhi ʻōniʻo).

paka probably represents the greatest potential danger because it is so numerous. A writer on Lanai had this to say of the *pūhi paka* one hundred years ago: "He often baffles the efforts of the fishermen. He will swallow the hook and bite the line in two. He will force himself out of a net, and if you have got him with a stout hook and line you must tear him to pieces before you can drag him out of the hole in the rocks in which he has braced himself. . . he will take off a toe, or snap off an exposed naked foot if he gets a chance. Where he is found no crabs or little fishes are to be seen. . . . He devours everything " (2). This report describes accurately the difficulties of catching *pūhi paka*, but exaggerates its destructive potential. Because the needlelike teeth of most morays are adapted for grasping prey, not cutting, it is unlikely that even the largest moray could "snap off" a human foot. Usually the moray that strikes a human hand or foot does so in error. Most such instances occur when an unseen *pūhi* strikes the hand of a diver who has reached back into a crevice for a lobster or an attractive shell. The moray in this situation probably feels threatened or mistakes the hand for prey. Generally, it will release its hold as soon as the error is recognized, and if the diver can resist his nat-

ural impulse to pull free, he may well escape with no more than a series of puncture wounds. Unfortunately, such presence of mind in this situation is rare, and a hand is often severely slashed as it is forcibly wrenched from between the backward-projecting teeth of the moray.

The larger morays are primarily fish-eaters, and their fanglike teeth are adapted for grasping mobile prey. Although most members of the family have this type of dentition, morays in the genus *Echidna*, of which three species are known to occur in Hawaii, possess instead blunt pebble-like teeth that are adapted for crushing. The prey of these blunt-snouted *pūhi*, which grow to about 3 feet long, are mostly large, heavily shelled crustaceans.

Not all the prominent eels on the reef are morays. One of the best-known Hawaiian species is the white eel, *Conger marginatus* (Plate 5), of the family Congridae. This fish, called *pūhi ūhā* by Islanders, grows to about 4 feet long, but is not a threat to humans. It remains secreted in caves during the day, but searches for prey in the open at night. The *pūhi ūhā* has always been a favorite food of Hawaiians; one early writer described the white eel as ". . . a fish of which chiefs were fond . . . considered choicer than wives . . ." (2).

Plate 4. Moray eel, *Gymnothorax flavimarginatus* (pūhi paka).

Plate 5. White eel, *Conger marginatus (pūhi ūhā)*.

Lizardfishes

The lizardfishes, family Synodontidae, collectively called 'ulae by Hawaiians, are well named. Their appearance is reptilian, both in body form and also in the way they rest motionless, but alert, on the sea floor. Usually these predators sit on a patch of sand among the corals, where their cryptic coloration makes them difficult to see. In this pose, fully exposed, yet effectively concealed, they wait for small fishes to come within striking range. Sometimes they bury themselves—all except their eyes and the tips of their snouts—in the sand. From this position, attacking with an explosive rush upward, they seize even the most wary prey between gaping, fang-rimmed jaws.

The four lizardfish most often seen on the reef—one species of *Saurida* (Plate 6) and three species of *Synodus*—are very similar in both appearance and habits. Most of them are less than 12 inches long (10).

Trumpetfish and Cornetfish

The trumpetfishes, family Aulostomidae, are represented in Hawaii by one species—*Aulostomus chinensis* (Plate 7), called *nūnū* by Hawaiians. The very similar cornetfishes, family Fistulariidae, are also represented in the Islands by just one species—*Fistularia petimba*. Both of these elongate fishes, with their long flute-shaped snouts, are distributed widely through warm waters of the Indian and Pacific oceans. The anatomies of the two differ only relatively little: perhaps the most recognizable difference lies in the long filament that streams from the center of the cornetfish's tail, a feature the trumpetfish lacks. The colors of each are more distinguishing: three color varieties occur among trumpetfish—plain brown, brown with lighter markings, and plain yellow; in contrast, the cornetfish is always pale green with lighter markings. There is also a difference in size: whereas the trumpetfish does not exceed a length of about 2 feet, the cornetfish often grows to 5 feet long, and more (10).

Both species are predators. The trumpetfish regularly captures both crustaceans and fishes, but the larger cornetfish is almost exclusively a fish-eater. On casual glance, the long tubular snouts of these fish suggest diets of very small prey, but in fact the snouts are capable of much expansion, and surprisingly large prey are captured. The expansion is so sudden that a strong vacuum is created within that literally sucks prey into their mouths. In fact, the mechanism so effectively snares large prey that it occasionally leads to disaster when one of these predators chokes to death on an oversized organism that has become tightly wedged in its throat.

Their feeding mechanism is effective only at

16

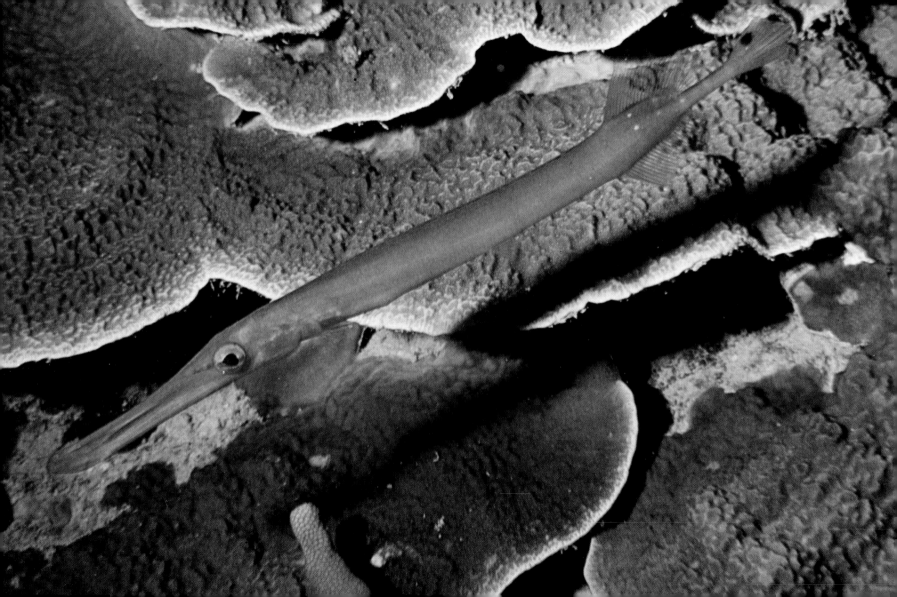

short range, however, and because neither species is a strong swimmer they must rely on stealth to get close to their quarry. Sometimes they approach prey from behind the body of a large animal that grazes on seaweeds, such as a parrotfish. This tactic is especially effective because many smaller fishes, like wrasses, characteristically assemble where herbivores are grazing. The smaller fishes congregate here and capture organisms uncovered when the seaweeds are cropped from the ocean floor. But the attraction can be fatal, as these smaller fishes sometimes fall prey to a trumpetfish or cornetfish lurking in the shadow of the big herbivore. Nevertheless, despite the frequency of such attacks, they capture most of their prey by quietly gliding along in the shadows of the reef, striking creatures that have carelessly strayed too far from cover (13, 14).

Squirrelfishes

The squirrelfishes, family Holocentridae, are prominent on Hawaiian reefs. Most are one or the other of two general types: species of one type are known collectively by the Hawaiian name *'ū'ū*, or probably more often today by the Japanese equivalent *menpachi;* species of the other type are known collectively by the Hawaiian name *'ala'ihi.* The coloration of all is predominantly red, and although some occasionally exceed a length of 12 inches, most are considerably smaller than this.

Including three very similar species that generally are not distinguished by Islanders, the *menpachi* are numerous on shallow reefs. Science classifies these as *Myripristis argyromus* (Plate 8), *M. berndti,* and *M. multiradiatus.* They are among the fishes most frequently hunted by spear fishermen, and, being highly regarded as food, bring a high price in Hawaiian markets today. *Menpachi* are nocturnal, a fact consistent with their very large eyes. They assemble in submarine caves during the day, but at night venture out and hunt small prey in the darkened waters over the reef.

Squirrelfishes that are called *'ala'ihi* by Hawaiians are classified today in the genus *Holocentrus.* Several forms were distinguished by early Islanders, including *'ala'ihi piliko'a ('ala'ihi* clinging to coral), *'ala'ihi kalaloa ('ala'ihi* with long spike), and *'ala'ihi lākea ('ala'ihi* with white dorsal fin). It is uncertain how these specific names relate to the six species of *Holocentrus* recognized today as common on shallow reefs: *Holocentrus diadema, H. lacteoguttatus, H. sammara, H. spinifer, H. tiere,* and *H. xantherythrus* (Plate 9). In contrast to the highly palatable *menpachi,*

Plate 8. Squirrelfish, *Myripristis argyromus* (*'ū'ū,* or *menpachi).*

Plate 9. Squirrelfish, *Holocentrus xantherythrus* ('ala'ihi).

the various 'ala'ihi are not used much as food today (10); nevertheless, it is said that 'ala'ihi were the favorite fish of King Kamehameha III (2). Similar to *menpachi*, 'ala'ihi are nocturnal animals that shelter themselves in reef crevices by day, and forage in the open under cover of darkness.

Silver Perch

The silver perch, called *āholehole* by Islanders, is *Kuhlia sandvicensis*, of the family Kuhliidae. This silvery fish with large eyes grows to a maximum length of about 12 inches. The young are numerous in tide pools, but larger individuals live offshore, mostly in water less than 20 feet deep. The adults, at least, secrete themselves in caves during the day; they emerge and feed at night, and at this time frequently are caught by fishermen (10).

The *āholehole* was a favorite of early Hawaiians, both as food and as a ceremonial sacrifice. Many chiefs valued the delicate flavor of this fish. Royalty at Hilo were known to have *āholehole*, still living, brought to them from the fishing grounds at Puna, over 20 miles away.

To keep the fish alive during this long overland journey, the natives wrapped them in seaweed (2). Ceremonial roles of the *āholehole* were several. It was regarded as a "sea pig" (*pua'a kai*), and as such could be substituted in ceremonies calling for a pig when this animal was not available. It was also offered in ceremonies requiring a white fish. An alternate meaning of the word *āholehole* led to still another ceremonial use. *Hole*, a component of the word, means "to strip away," and so this fish was sacrificed when the object was to dispel evil spirits (2). In the early days of immigration to Hawaii, Caucasians sometimes were called *āholehole* because of their white skins (2).

Bigeyes

Of the bigeyes, family Priacanthidae, several species reportedly frequent Hawaiian waters; nevertheless, only one is numerous near shore: *Priacanthus cruentatus* (Plate 10). Known to Hawaiians as the *'āweoweo*, this fish has long been a reef species much sought after by Island fishermen. Growing to about 12 inches long, the *'āweoweo* is widespread on reefs throughout

Plate 10. Bigeye, *Priacanthus cruentatus* ('āweoweo).

Plate 11. Cardinalfish, *Apogon menesemus* ('*upāpalu*).

tropical seas. It can rapidly change its coloration from a reddish to a silvery hue, or vice versa; sometimes it assumes a blotched pattern composed of these two hues in combination. Like the *menpachi*, *'ala'ihi*, and *āholehole*, the *'āweoweo* remains in the reef's shadows during daylight, and hunts prey in open water only after dark (10).

Occasionally great schools of *'āweoweo* appear near shore at night. Early Hawaiians witnessed this spectacular event with mixed emotions. Although pleased with the sudden abundance of an esteemed food fish, the coming of these immense schools also evoked sadness and awe, portending as they did the imminent death of a high chief (2).

Cardinalfishes

Seven species of cardinalfishes, family Apogonidae, are abundant on Hawaiian reefs; however, only two are frequently seen by the casual observer: *Apogon menesemus* (Plate 11) and *A. snyderi*. These two are very similar, and both carry the Hawaiian name *'upāpalu*. The *'upāpalu* are the largest cardinalfishes in Hawaii, growing to about 9 inches long (10). Although both species of *'upāpalu*, like the other cardinalfishes, seek cover by day, individuals of *A. snyderi* frequently hover at the openings to their shadowy retreats and here are often seen. The various other species, all smaller fishes, are more secretive, and only rarely are visible in daylight. Like so many fishes that frequent caves in the reef by day, the cardinalfishes emerge into the open and hunt prey after dark. *'Upāpalu* are caught readily by fishermen at night, especially under moonlight, and this fact has lead many Islanders to call them "moonlight Annies."

Jacks

The jacks, family Carangidae, include many game fishes that are the favorites of Island sportsmen who cast their lines from rocky shores or use a spear underwater. As is true of many other fishes mentioned above, it remains uncertain how some of the names for jacks used by early Hawaiians relate to species recognized today. There is no question, however, that the jacks have always been important to Islanders for both food and sport.

Species of the genera *Caranx*, *Carangoides*, and *Gnathanodon* are among the largest and most sought after by fishermen. To Hawaiians these are the various kinds of *ulua*. The young of most are collectively called *pāpio*, but the adults generally are distinguished from one another.

The blue *ulua*, also called *'ōmilu*, is *Caranx melampygus* (Plate 12). This is the most frequently seen of the larger jacks on the reef, and attains a length of 3 feet. *C. sexfasciatus*, called *pake ulua*, is probably the largest of the jacks taken by sportsmen, sometimes growing to over 5 feet long (10).

Several other species of *ulua* are similar in form and habits to the above, but *Gnathanodon speciosus*, called *pa'opa'o ulua*, has habits that set it apart from other jacks on the reef. Whereas the others are mostly strong swimmers that run down their prey in open water, the *pa'opa'o ulua* looks much like a grazing herbivore as it feeds head down with mouth thrust amid vegetation that carpets the sea floor in some areas. Despite appearances, this fish is strictly a predator, working its highly protrusible, toothless jaws through the seaweeds to capture tiny animals hidden there (13). This is the habit only of adults, which may attain a length of 3 feet. The very young of this species, when only a fraction of an inch long, swim close among the venomous tentacles of jellyfishes, a habit that offers them protection from predators.

When they become too large for this habit, but still are less than an inch long, the juveniles become "pilots," swimming close before the snouts of much larger fishes, usually predators. Because the eyes of most of these big predators are on the sides of their heads, they cannot see the area immediately in front of their snouts. It is here that the tiny jacks swim—within inches of the predator's jaws, but out of sight and so probably unnoticed. Not only are the young jacks relatively safe from other predators when they swim immediately in front of their large companions, but also they are literally pushed along in the big animal's bow wave. By hitch-hiking in this way, the little jacks travel distances they could not possibly cover on their own. Thus this habit not only provides these little fish a haven from predators, but also aids in their distribution. Sometimes, mistaking a human skin diver for a large fish, tiny *pa'opa'o ulua* take up a station directly in front of the diver's face mask. These jacks can maintain the piloting habit until they reach 5 or 6 inches long by swimming with progressively larger fishes (15).

Goatfishes

The goatfishes, family Mullidae, are prominent members of the reef community, and also have been important to Hawaiian fishermen since early times. Most of them grow to about 16 inches long, and all possess a pair of large barbels under the chin, a characteristic they

Plate 12. Blue jack,
Caranx melampygus
(*'ōmilu,* or *ulua).*

Plate 13. Goatfish, *Parupeneus porphyreus (kūmū).*

Plate 14. Goatfish, *Mulloidichthys samoensis (weke 'a'ā)*.

alone possess among nearshore reef fishes. With these barbels, the goatfishes probe the sand and vegetation of the ocean floor for small organisms concealed there.

The *kūmū, Parupeneus porphyreus* (Plate 13), has always been a great favorite of Hawaiians. This red goatfish was important to early Islanders, both as a food fish and as a ceremonial sacrifice. It was one of the "sea pigs," as discussed above, and also was used when a *kahuna* demanded a red fish for ceremonial sacrifices.

Several other species of the genus *Parupeneus* are common, including *P. pleurostigma,* called *malu* (10); *P. chryserydros,* called *moano kea* (10); *P. bifasciatus,* called *munu;* and *P. multifasciatus,* called *moano.* The *moano,* especially, has long been highly esteemed, as reflected in the old chant "... *A he moano ka lena, Ono! Ono!"* which means "The *moano* of the yellowish sea, Delicious! Delicious!" (2).

Several different goatfishes comprise the various species that Hawaiians call *weke. Weke 'a'ā* (staring *weke*), *Mulloidichthys samoensis* (Plate 14), is probably the most numerous *weke* on nearshore reefs. Usually members of this species occur in schools that hover close to the sea floor, or as individuals or in groups of two or three that probe with their barbels for prey in sand patches on the reef.

Weke 'a'ā and, more often, *weke pahulu* (nightmare *weke*), *Upeneus arge,* occasionally produce

hallucinations, or morbid nightmares and acute depression in people who have eaten them. Although many people claim that the affliction occurs only after eating the brains of these *weke,* the symptoms also have appeared in those who have consumed only the flesh of the body. In Hawaii during recent times the phenomenon has been limited to the islands of Kauai and Molokai, and then only during the months of June, July, and August (16). Nevertheless, the nightmare-inducing quality of certain *weke* was general knowledge among early Islanders, suggesting that the malady was more widespread in times past. Legend places the origin of the affliction on the island of Lanai, where it is related to the death of Pahulu, chief of evil spirits. As one story goes, Pahulu leaned over a pool of water, and was struck by a stone hurled down on him by an adversary concealed in a tree above. Pahulu fell lifeless into the water, but his spirit survived, embodied in the *weke pahulu;* thus, contends the tale, people who eat this fish are troubled with nightmares (17).

Whereas certain *weke,* such as *weke 'a'ā,* are light-colored, others, like *weke 'ula* (red *weke*), *Mulloidichthys auriflamma,* are reddish. Thus, in ceremonial offerings calling for a red fish, or a white fish, an appropriate *weke* species could be served. The meaning of the word *weke,* "to open," also brought these animals into use as a sacrifice when a *kahuna* wished to "open up" the mind and release evil thoughts (2).

Nenue is the Hawaiian name for the rudder-fish, *Kyphosus cinerascens* (Plate 15), which is the only member of the family Kyphosidae common on Island reefs. They grow to lengths of about 2 feet. Almost all *nenue* are grey, although patterns of light and dark are variable. A few individuals, however, carry irregular blotches of yellow, and rarely one is yellow all over. Early Hawaiians believed these aberrantly hued individuals to be protectors of other *nenue*, and referred to them as the *makua* (2).

Nenue feed on seaweeds, and are not regarded as a food fish by most people today.

Their strong odor is repugnant to modern tastes, but in the old days this same characteristic made them especially desirable. In ancient Hawaii *nenue* were in such high demand that the supply often was reserved for the chiefs (2). One can scarcely find a better example of the extent to which tastes have changed. A favorite relish was prepared by chopping up into small pieces the head of a *nenue*, adding this to the fish's entrails, which have an especially penetrating odor of their own, and then seasoning the mixture with *kukui* nuts and chili pepper (2).

Rudderfish

The family Chaetodontidae, the butterflyfishes, is probably more characteristic of coral reefs than is any other major family of fishes. If Hawaiians distinguished most of the varied butterflyfishes in their waters, little evidence of this remains today. Perhaps they felt no need to differentiate one from another; after all, these fishes, which seldom exceed a length of about 8 inches, were little valued as food. Most seem to have been called, collectively, *lau hau* (*hau* leaf), or *kīkākapu* (*kīkā*, strong; *kapu*, taboo). This second name suggests that butterflyfishes had religious significance, as do references to them in such old chants as: "*He kākau kī oki 'ōni o i ka lae he kī 'oki 'o ke kīkākapu 'o ka i 'a kapu,*" which means,

"Marked with bars and streaks on the forehead, the *kīkākapu* is a sacred fish" (18).

Butterflyfishes bring to Hawaiian reefs brilliant colors arranged in bold patterns; indeed, this is their single most outstanding feature. Because words fail, portrayal of these colorations is left to photographs (Plates 16–25). Much speculation circulates as to the function of these bright hues. Probably the view most often expressed is that the intricate patterns actually camouflage the fish as it swims amid the riot of colors and shapes comprising a coral reef. However, one who has actually viewed butterflyfishes underwater on Hawaiian reefs suspects this to be the view of an armchair theorist.

Butterflyfishes

31

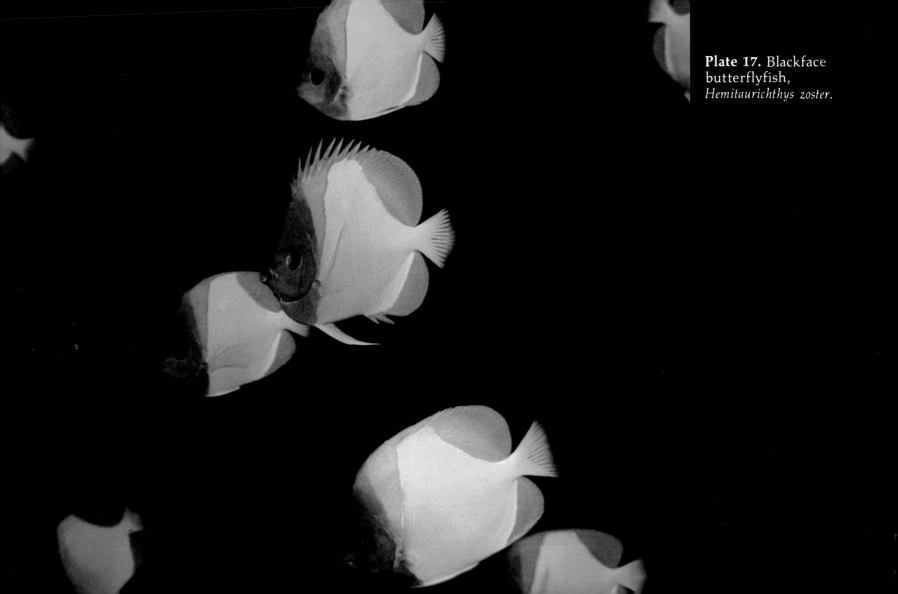

1, *Centropyge potteri.*

Plate 22. Pebbled butterflyfish, *Chaetodon multicinctus.*

Plate 23. Ornated butterflyfish, *Chaetodon ornatissimus.*

Plate 24. Four-spot butterflyfish, *Chaetodon quadrimaculatus.*

The fact is that these fishes are conspicuous in their natural habitat. Some elements of their color patterns seem to conceal certain body features, the most notable being the black bars that effectively conceal the eyes of so many species; nevertheless, it seems that the overall effect is one of display. It does not seem that camouflage should be especially important for butterflyfishes. Their prey are mostly sedentary invertebrates, so there is no need for them to conceal themselves from their prey. Furthermore, they do not swim far from the coral, so that if they themselves are threatened by other predators shelter is never more than a quick dart away. Possibly the butterflyfishes' distinctive colorations advertise the fact that they are a risky and unrewarding meal. There is very little meat on these deep-bodied fishes, and with many possessing strong fin spines it would not seem worthwhile for a predator to chance getting one lodged in its throat (19). Recognition of other members of their own species may be an overriding need of butterflyfishes, especially as most characteristically occur in pairs, and coloration may be related to this. Hawaiian reefs are populated by a wide variety of butterflyfishes, with no one species being especially numerous. One regularly sees eleven different species of the genus *Chaetodon*, with several others appearing on occasion; of the genus *Centropyge*, one species is common, three others are seen just now and then; the genus *Hemitaurichthys* is repre-

sented by two species, as is the genus *Forcipiger* (see below); finally, there is a single species of the genus *Holacanthus*. In such an assemblage, it would seem that members of the various species, all similarly shaped, would need ready means of sorting themselves out.

All of the butterflyfishes have flexible, comblike teeth, but a wide variety of feeding habits are represented. Diets range from plankton and small bottom-dwelling shrimps to sponges; from worms and sea slugs to corals. Although the food of many butterflyfish species is restricted to a narrow range of organisms, others, such as *Chaetondon auriga*, have varied diets. This last group—those taking assorted foods—include the ones called "steal bait" by fishermen. This name is based on their habit of nibbling from the hook whatever bait the fisherman might be using.

The first species of fish from Hawaii to be recognized by science is the butterflyfish *Forcipiger longirostris*. This species is very similar to the other *Forcipiger* in Hawaii, *F. flavissimus* (Plate 26). Because of their elongated snouts, the two are distinct from all other Hawaiian butterflyfishes. No doubt this led early Hawaiians to distinguish them from the others as *lau wiliwili nukunuku 'oi'oi* (sharp-beaked *wiliwili* leaf).

Forcipiger longirostris was first described in 1782, based on a specimen collected in the Hawaiian Islands during Captain Cook's third

voyage in H.M.S. *Resolution* (1776–1780). Here unfolds a story that illustrates the sort of problems ichthyologists face in trying to classify the vast array of fishes that inhabit tropical reefs.

It was long believed that just one species of *Forcipiger* inhabits Hawaiian waters, and this was assumed to be the form collected by Captain Cook. Individuals of this common species are numerous elsewhere in the Pacific and were also assumed to be *F. longirostris*. Finally, in 1961, Dr. John E. Randall, ichthyologist at the Bernice P. Bishop Museum, reported a second relatively rare species of *Forcipiger* from the Kona coast of the island of Hawaii. Believing this to be an undescribed species, Dr. Randall named it *Forcipiger inornatus* (20). Then, in 1964, Dr. Alwyne Wheeler, of the British Museum, took a new look at Captain Cook's original specimen, and to his surprise found it to be different from the common form so long known as *F. longirostris* (21). As chance would have it, Captain Cook, who spent much time at Kealakekua Bay, Kona, had collected the rare form of *Forcipiger* that was rediscovered by Dr. Randall almost 180 years later. The rare form, then, is the true *F. longirostris*, whereas the common form, so long incorrectly recognized by this name, is the species now named *F. flavissimus*. Where did the name *flavissimus* come from? This name had been given long ago to representatives of this species in Mexican waters. It was no longer used after the Mexican form was recognized to be the same as that common in Hawaii and the western Pacific. *Flavissimus*, being the second oldest name given to this species, became the valid name when it was discovered that *longirostris* no longer applied.

Further studies by Dr. Randall have shown that, unlike members of *F. flavissimus*, which are all similarly hued, members of *F. longirostris* may have either one of two distinct colorations. The coloration of some is essentially the same as *F. flavissimus*; it was one of these that was taken by Captain Cook's expedition, hence the long-standing confusion. Others are a solid dark brown; Dr. Randall's description of *F. inornatus* is based on one of these.

The hawkfishes, family Cirrhitidae, include several colorful species that perch themselves in plain view on the coral during the day. These fishes are known by Hawaiians appropriately, as *pili ko'a* (coral clinging), and include *Paracirrhites forsteri* (Plate 27), *P. arcatus* (Plate 28), and *Cirrhitops fasciatus* (Plate 29). All are relatively small, most of them less than 6 inches long. They capture small prey by darting suddenly from their resting spots on the coral, and this

Hawkfishes

Plate 27. Hawkfish, *Paracirrhites forsteri (pili koʻa)*.

acirrhites arcatus (pili koʻa).

habit presents a paradox. Usually predators that feed this way are colored much like their backgrounds; this cryptic coloration allows them to go unnoticed by their prey, which are thus more likely to stray within striking range. How, then, does one account for the conspicuous coloration of the *pili ko'a*?

Another type of hawkfish prominent on Hawaiian reefs is *Cirrhites pinnulatus* (Plate 30), called by Hawaiians *po'o pa'a* (stubborn). Unlike the *pili ko'a*, the *po'o pa'a* spends most of its time during the day hidden among the coral. It is a nocturnal fish that is larger than the *pili ko'a*, attaining a length of about 10 inches. Probably because of its larger size, the *po'o pa'a* alone among the hawkfishes has long been considered important as food.

Damselfishes

The damselfishes, family Pomacentridae, are numerous on Hawaiian reefs, but they do not seem to be nor to have been important to Islanders. In fact, so far as is known today, most do not have Hawaiian names. Probably this is because most are less than 6 inches long when fully grown, and thus are too small to be significant as food.

Members of six species feed on tiny crustaceans in the plankton, and can be seen during the day in aggregations that hover several feet above the reef. These include *Dascyllus albisella*, *Abudefduf abdominalis* (called *maomao* by Hawaiians), and four species of the genus *Chromis*: *C. leucurus* (Plate 31), *C. ovalis*, *C. vanderbilti*, and *C. verater*. When threatened, the fish in these aggregations close ranks and dive to shelter on the reef below them.

Members of five other species are scattered about close to the reef, where they feed on various bottom-dwelling organisms. One of these is *Pomacentrus jenkinsi* (Plate 32), which is one of the most widespread and numerous fish on Hawaiian reefs. Others include *Plectroglyphidodon johnstonianus* (Plate 33), *Abudefduf sindonis*, *A. imparipennis* (Plate 34), and *A. sordidus*. Juveniles of this last species, called *kūpīpī* by Hawaiians, live in tide pools. Being extremely aggressive, they dominate most other fishes in their pools. As they mature, *kūpīpī* leave the tide pools for deeper water in surge channels and on the reef-face. When fully grown, *kūpīpī* are probably the largest of the damselfishes, attaining lengths of up to 10 inches. Probably for this reason, it is the *kūpīpī*, among damselfishes, that is most often mentioned as a food of early Islanders (22).

The wrasses, family Labridae, number more species on Hawaiian reefs than does any other family of fishes, even though there are considerably fewer than had been thought until very recently. They probably number a few over thirty, which is less by over one-third the number that were recognized as recently as ten years ago. The difficulty is that wrasse species generally include two or more distinct forms: coloration, especially, differs with age and sex. To further complicate matters, it is widespread in this family, perhaps universal, that individuals change sex at some point during their lives (23). For these reasons, many species contain two or more distinct forms that long had been thought distinct species themselves. Even the present estimate of the number of species is tenuous, however, as biological sleuthing continues to unravel the true relationships among the different forms.

The colorful wrasses are among the more distinctive fishes on Hawaiian reefs. They include fishes of all shapes and sizes, although most are less than a foot long. Usually they swim with a characteristic rowing motion of their pectoral fins; the sweeping tail strokes that propel most other fishes are used by wrasses only when sudden acceleration is needed. Wrasses have small mouths, with sharp, pointed teeth, and carry in their throats heavy bones that crush the shelled animals upon which most of them feed. All the wrasses restrict their activity to daylight; after dark they rest under cover on the reef or under the sand (10).

The ubiquitous wrasses have long been well known to Islanders. A number of the more prominent species are known to Hawaiians as varieties of *hīnālea*. *Thalassoma duperreyi* (Plate 35), called *hīnālea lauwili*, occurs only in Hawaiian waters; it is probably the most numerous of the commonly observed fishes in the various reef habitats. Other species recognized as varieties of *hīnālea* are *Thalassoma ballieui* (Plate 36), called *hīnālea luahine* (old woman *hīnālea*); and *Gomphosus varius* (Plate 37), called *hīnālea nuku 'i'iwi* (beak-like-a-bird *hīnālea*). Several other varieties of *hīnālea* were recognized, but have not yet been identified with species known today.

Hawaiians often ate *hīnālea* when drinking *'awa*, considering them to provide a good aftertaste. For this purpose, they commonly kept the *hīnālea* in pools, so as to be handy when needed (2). *Hīnālea* were the major ingredient of *i'a ho'omelu*, a favorite dish. For this dish, the fish was allowed to partially decompose before being seasoned with *kukui* nuts and chili peppers. Not surprisingly, the preparation had a strong, foul odor, which led to the phrase *ipu kai hīnālea* (dish of *hīnālea* sauce) being applied to someone with unpleasant breath (18).

Hīnālea had religious significance. In one ceremony, these fishes were offered to the gods as an aid to bring on pregnancy (2). The *hīnālea nuku 'i'iwi* was used by *kāhuna* as a *pani*, which is

Plate 35. Saddleback wrasse, *Thalassoma duperreyi* (*hīnālea lauwili*).

Plate 39. Hogfish, *Bodianus bilunulatus* ('a 'awa).

Plate 40. Eight-lined wrasse, *Pseudocheilinus octotaenia*.

Plate 41. Cleaner wrasse, *Labroides phthirophagus.*

a bit of food given to close a period of medical treatment. The affliction treated was mental illness, and this particular fish was appropriate because a second name for the species, *hīnālea 'aki lolo*, is pertinent (*'aki*, to nibble; *lolo*, brain) (2).

Other prominent wrasses include *Halichoeres ornatissimus* (Plate 38), called *lā'ō* (sugarcane leaf); *Anampses cuvieri*, called *'ōpule*; *Bodianus bilunulatus* (Plate 39), called *'a 'awa*; *Cheilio inermis*, called *kūpou*; *Hemipteronotus taeniourus*; and *Pseudocheilinus octotaenia* (Plate 40). Truly this is a varied group.

Several gaily hued forms of the genus *Coris* were recognized by ancient Hawaiians as varieties of *hilu*, a word meaning well behaved. It was said that women who craved this fish during their pregnancy would give birth to quiet, dignified children (2). Certain *hilu* were considered gods who could assume human form at will (2).

One little wrasse that is not known to have attracted the attention of ancient Hawaiians has excited the interest of some present-day Islanders. This is *Labroides phthirophagus* (Plate 41), which does not exceed a length of more than a few inches, and which has the habit of picking external parasites off the bodies of other fishes. These wrasses are not numerous on the reef. Generally only a few of them inhabit a given area, and these individuals occur at well-defined locations called "cleaning stations." The other fishes on the reef know the location of these stations, and swim there when they need their bodies cleansed of parasites. Upon arriving at the cleaning stations, fishes pose motionless in a characteristic fashion. The soliciting attitude of these fishes attracts the cleaner, which approaches them, inspects their bodies, and then may proceed to pick off the irritating parasites. Sometimes many fishes hover together at a cleaning station, all waiting to have their parasites removed by the little cleaner (24).

Parrotfishes

The colorful parrotfishes, called *uhu* by Hawaiians, are another group especially characteristic of coral reefs. They are grouped in a family called the Scaridae. Parrotfishes are closely related to the wrasses, which they strongly resemble. Generally they are larger than wrasses, however, with members of some species growing to about 2 feet long. Perhaps their most distinctive feature is a parrotlike beak, formed by the fusion of their teeth. This feature, combined with the brilliant hues that many possess, makes the parrotfishes a well-named group.

With their heavy beak, parrotfishes scrape algae off the surfaces of rocks and dead coral, leaving characteristic marks on these surfaces. In this activity, parrotfishes also scrape away

63

much of the coral rock itself. After this material is ground to a fine powder by bones in the parrotfish's throat, it passes on through the digestive tract, and finally is deposited on the sea floor. It has been suggested that much of the fine coral sand on a coral reef has passed through the guts of parrotfishes (25). Parrotfishes seem to be defecating much of the time, a fact probably leading to the Hawaiian name of one form, *uhu pālukaluka* (*uhu* with looseness of the bowels).

Parrotfishes are active only in daylight; at night they rest among the rocks and coral on the reef. At this time many of the smaller individuals secrete a sheath of mucus about themselves (Plate 42), presumably as defense against nocturnal predators (26).

Uhu have long been favorites of Islanders. Many Hawaiian names refer to different forms of *uhu*, but are difficult to associate with species known today. Even the names scientists use for several forms of Hawaiian parrotfishes remain uncertain. The main problem is that, just as with the wrasses, a given parrotfish species usually has several forms, with distinctions being associated with age and sex. Also, as with wrasses, the situation with *uhu* is further clouded by the fact that some individuals, perhaps all of them, change their sex at some time during their lives (27).

Some distinctive forms nevertheless seem identifiable among the names used by early Islanders. *Uhu 'āhiuhiu* (untamed *uhu*), described as a large blue species (2), is probably the large male of *Scarus perspicillatus*, but could possibly be the male of *Calotomus spinidens* (Plate 43); *uhu pālukaluka*, described as reddish brown on its upper body, and bright red on its lower (2), is probably the adult female of *Scarus rubroviolaceus*; and *uhu 'ele'ele* (dark *uhu*), described as a large green fish (2), may be the adult male *Scarus rubroviolaceus* (Plate 44). Among Island parrotfishes for which Hawaiian names cannot be suggested is *Scarus dubius* (Plate 45).

The prominence of *uhu* on Hawaiian reefs is reflected in their frequent reference in Island lore. One legendary *uhu* was the fish-god Uhumāka 'īkā'i (roving, sight-seeing *uhu*). It is told that when very young, Uhumāka 'īkā'i was caught by a fisherman who raised him to great size in captivity. When fully grown, Uhumāka 'īkā'i was released to the sea, but thereafter answered his master's bidding to drive vast numbers of fishes up onto the beach. In this way, the people living on the shore between Makapu'u and Kualoa on Oahu were regularly supplied with a superabundance of food (28). Despite this service to the people, Uhumāka 'īkā'i was hunted by the legendary hero Kawelo. Taking to sea in a canoe, Kawelo sought his quarry for two days off the Waianae coast. Finally Uhumāka 'īkā'i appeared, his approach heralded by gathering storm clouds. Kawelo threw his net, entrapping the mighty fish-god, and a

Plate 44. Parrotfish, *Scarus rubroviolaceus (uhu),* adult male.

Plate 45. Parrotfish, *Scarus dubius (uhu)*, adult male.

wild battle between powerful antagonists ensued. Uhumāka 'īkā'i could not free himself from the net. Kawelo could not bring him in, and could only hang on as the canoe was towed far from land. The struggle continued north as far as the island of Kauai, then back to Oahu. At last, with help from other Islanders, Kawelo subdued and then killed the great *uhu* at Waikīkī (29).

In the old days, fishermen believed that the way they saw *uhu* behaving at sea told them how their wives were behaving at home. Thus, when *uhu* frolicked in the water, a fisherman knew that his wife was guilty of improper levity at home. If two *uhu* were seen rubbing their snouts together, the fisherman knew it was time to stow his gear and return home, where an unfaithful wife needed punishing (2).

Surgeonfishes

Surgeonfishes, which comprise the family Acanthuridae, are especially numerous in Hawaii. In fact, probably more of the fishes one sees swimming over Island reefs belong to this family than to any other. These colorful animals, which range in length from about 7 inches to over 2 feet, owe their name to the knifelike spines that each one carries on the base of its tail. Most surgeonfishes have a single sharp spine on each side that folds into a sheath of skin; the fish erects these spines when threatened, and with them can seriously injure an unwary fisherman. Other surgeonfishes carry on each side a number of fixed spines, which are blunt in comparison to the single, folding type.

The various different Hawaiian reef habitats each have a characteristic assemblage of surgeonfish species. Where waves crash against the reef, and the water often is a swirl of bubbles and turbulence, the surgeonfishes present usu-

ally include *Acanthurus achilles* (Plate 46), called by Hawaiians *pāku'iku'i*; *A. guttatus*; and *A. leucopareius*, called *māikoiko*. On the other hand, surgeonfishes associated with a sea floor richly overgrown with coral, in deeper water, usually include: *Acanthurus sandvicensis* (Plate 47), called *manini*; *A. nigrofuscus*; *A. nigroris*, called *maiko*; *Ctenochaetus strigosus* (Plate 48), called *kole*; *Zebrasoma flavescens* (Plate 49), called *lau'īpala*; and *Naso lituratus* (Plate 50), called *kala*. When the reefs in water deeper than about 30 feet are interspersed with patches of sand, the characteristic surgeonfishes include *Acanthurus dussumieri* (Plate 51), called *palani*; *A. xanthopterus*, called *pualu*; and *A. olivaceus* (Plate 52), called *na'ena'e*. All those mentioned so far are bottom feeders; most, like the *manini*, crop seaweeds from a rocky substrate (Plate 47), but others, like the *na'ena'e*, sift through sand for tiny food items that have accumulated there (Plate 52). A few sur-

Plate 49. Surgeonfish, *Zebrasoma flavescens (lau ʻīpala)*.

Plate 50. Surgeonfish, *Naso lituratus (kala)*, lying on the coral at night, with a small unidentified shrimp, perhaps a cleaner, on its lower jaw.

Plate 53. Surgeonfish,
Naso hexacanthus (kala).

geonfishes find their food up in the water above the reef: these include *Acanthurus thompsoni* and *Naso hexacanthus* (Plate 53), which feed mostly on tiny animals that drift as plankton in the mid-waters (30).

The surgeonfishes have long been a favorite food of Hawaiians, despite the fact that smaller species, like the *manini*, do not carry much flesh. Furthermore, many have a strong, unpleasant odor, especially some of the larger species, like the *palani*. An old Hawaiian tale recounts how the *palani* acquired its odor: Once, lost at sea, the legendary Ke'emalu remembered her ancestor among the surgeonfishes, a god named Palani nui mahao 'o. She called to him for aid, and the great fish soon appeared to carry her shoreward on his back. Midway through the journey, Ke-'emalu was overcome by a strong need to urinate. Unable to control herself, she urinated on her ancestor's back, an act which understandably infuriated the fish-god. Although he immediately threw Ke'emalu into the sea, leaving her to swim ashore on her own, the strong, unpleasant odor has persisted to the present day in this species of surgeonfish (2).

Moorish Idol

The moorish idol, *Zanclus canescens* (Plate 54), is closely related to the surgeonfishes, and some ichthyologists consider it to be a member of that family. However, it lacks the spines on the base of the tail fin, which are characteristic of surgeonfishes, and so is considered by other ichthyologists to represent a family of its own, the Zanclidae. It is generally regarded as one of the most graceful and distinctively beautiful species in the sea. It grows to about 9 inches long, and was called by Hawaiians *kihikihi* (angular). Usually swimming among the corals and rocks in groups of four to six or so individuals, the *kihikihi* probes for food with its long snout in the reef's crevices and narrow depressions.

Scorpionfishes

The scorpionfishes, family Scorpaenidae, owe their name to the venomous sting which they can inflict with their fin spines. These spines are deeply grooved, and within the grooves are venom glands (31).

Among representatives of the family in Hawaii, the lionfish, *Pterois sphex* (Plate 55), represents the greatest threat. This is only partly because its venom is especially virulent. In addition, a diver may find the temptation to pick one

up irresistable, because the species is one of the most ornate on Hawaiian reefs, and seemingly fearless. It is prized by aquarists, and even experienced collectors usually are stung at least once before learning to handle this fish with special care. Lionfish on Hawaiian reefs reportedly grow to a length of 10 inches (10), but those seen on the reef usually are less than half this size.

The closely related *Dendrochirus brachypterus* may have an equally virulent sting, but, even though it is more numerous on the reef than the lionfish, it probably is less a threat to humans. This is because the fish is a comparatively drab species, and thus not readily noticed nor sought after by collectors. *D. brachypterus* reportedly grows to a length of 7 inches (10), but again, those seen on the reef usually are smaller than this.

Two other prominent Hawaiian scorpion-fishes are *Scorpaenopsis gibbosa*, and *S. cacopsis* (Plate 56). These are the largest members of the family to be seen frequently on the reef,

with the latter attaining a length of 20 inches (10). The two are similar, and both are known by the Hawaiian name *nohu*. In contrast to the lionfish, which stands out so strikingly in its surroundings, the two *nohu* rest immobile and are difficult to distinguish because their colorations so closely match that of the reef. *Nohu* resemble the notorious stonefish, a denizen of islands far to the south and west of Hawaii. Fortunately for Hawaiian Islanders, the anatomical similarity of the *nohu* to the stonefish does not extend to the potency of its venom, which is mild by comparison. Despite their forbidding appearance, *nohu* have long been considered excellent eating.

The most numerous scorpionfishes on the reef are two species of the genus *Scorpaena*: *S. ballieui* and *S. coniorta* (Plate 57). Despite their large numbers, they are small fishes, usually under 4 inches long, and are not conspicuous. *S. coniorta* frequently is seen nestled among coral branches.

Gobies

The gobies, known to early Hawaiians as various forms of *'o'opu*, include an assortment of relatively small fishes, many less than an inch long even when fully grown. Sitting motionless as they do on a substrate colored much like themselves, they often go unnoticed by the casual observer. Some, the *'o'opu wai*, live in Hawaii's freshwater streams, and it is these that are most often mentioned in Island lore. Our concern here, however, are those living in the

Plate 56. Scorpionfish,
Scorpaenopsis cacopsis (nohu).

Plate 57. Scorpionfish, *Scorpaena coniorta*.

sea, the *'o'opu kai*. The saltwater gobies inhabit a variety of habitats, from deeper waters on the reef to tide pools along the shore.

Most of the gobies are classified in the family Gobiidae, an assemblage unique among Hawaiian fishes in that the pair of fins on their underside are fused together, forming a suck-ing disc. This structure effectively grasps the sea floor, and is particularly helpful to the fish in maintaining position where the water is turbulent. Probably the most abundant fish in Island tide pools is the goby *Bathygobius fuscus*, called *'ōhūne* (10).

Blennies

Hawaiian blennies, most of which were recognized by early Hawaiians as various forms of *pāo'o*, are grouped into two families. However, one of these, the family Tripterygiidae, is represented in the Islands by just one small, secretive species. The rest of the Hawaiian blennies represent the family Blenniidae. A major characteristic unifying these fishes is a pair of fins far forward under their throat. This feature serves well in propping them up on the substrate in their characteristic resting pose. They are elongated fishes, and swim with exaggerated undulations of the body. The largest of the blennies probably does not grow to over 7 inches long. They are among the more numerous fishes in shallow water and in tide pools along rocky shores.

The rockskipper, *Istiblennius zebra*, is probably the form Hawaiians called *pāo'o lēhei* (leaping *pāo'o*). This fish inhabits tide pools, and characteristically leaps from pool to pool in a way that indicates familiarity with the relative positions of the pools in its area (32). In one legend, a *pāo'o*, probably of this species, speaks to other fishes, describing its own activity in a chant:

This is *pāo'o, pāo'o*
That rogue, that mischief maker
That rests on the *līpoa* seaweed.
A nibble here, a nibble there.
I leap, I jump,
I leap into the large sea pools,
I leap into the small sea pools,
Poking this, taking that....(2)

Fishermen frequently took these blennies from the pools and used them for bait when casting their lines from rocky shores. In the process, they would often pop a live one into their mouths. These fish were not eaten raw once they had died, however, because they then have a bitter taste (2).

Although most of the blennies live close to shore, several species range out into the deeper parts of the reef. One of these, *Exallias brevis*

84

(Plate 58), called *pāo'o kauila*, has perhaps the most striking color pattern among Hawaiian blennies.

Most members of the family have moveable, comblike teeth with which they scrape algae and other nutritive materials from the surface of rocks and coral. However, two species that range into deeper water have exceptional feeding habits. These are the sabre-toothed blennies *Runula goslinei* and *R. ewaensis* (Plate 59). The sabre-toothed blennies further differ from others of their family in that they do not rest on the sea floor, but instead hover in the water above the reef. Blennies of the genus *Runula*, none of which grow to more than a few inches long, attack other fishes many times their own size. Launching their strikes from hovering positions above the reef, sabre-toothed blennies attack their quarry unseen from below or behind. Upon contact, the startled larger fish bolts forward and swims away, having lost to the blenny some of its body mucus and perhaps skin fragments. Sabre-toothed blennies owe their name to the pair of fanglike teeth they carry in their lower jaw. These teeth are not used in feeding, but perhaps serve some function in territorial defense. When not hovering above the reef, these blennies occupy abandoned mollusk shells or worm tubes embedded in the rocks or coral (13, 33).

Triggerfishes and Filefishes

Triggerfishes, family Balistidae, are distinctive, slow-moving creatures that appear deceptively harmless. The fact is, however, the relatively small mouths of most are equipped with exceptionally strong teeth and jaws that can seriously injure a careless fisherman. These features are adaptations to diets of hard-shelled organisms, like sea urchins.

Triggerfishes, most of which attain lengths of between 7 and 12 inches, owe their name to the mechanism that locks upright the long front spine of their dorsal fin. When this spine is thus positioned, it cannot be depressed until the smaller second spine, the trigger, is released. These fishes are served well by this structure when they are threatened. At such times they dive into narrow crevices in the reef, and wedge themselves into position by locking their spine erect. Thus secured, the fish cannot be removed from its refuge without breaking up the surrounding reef, or tearing up the fish.

To early Hawaiians, the different triggerfishes were various types of *humuhumu*. Two similar species, *Rhinecanthus rectangulus* and *R. aculeatus*, are well known in song as *humuhumu nukunuku a pua'a* (*humuhumu* with a snout like a

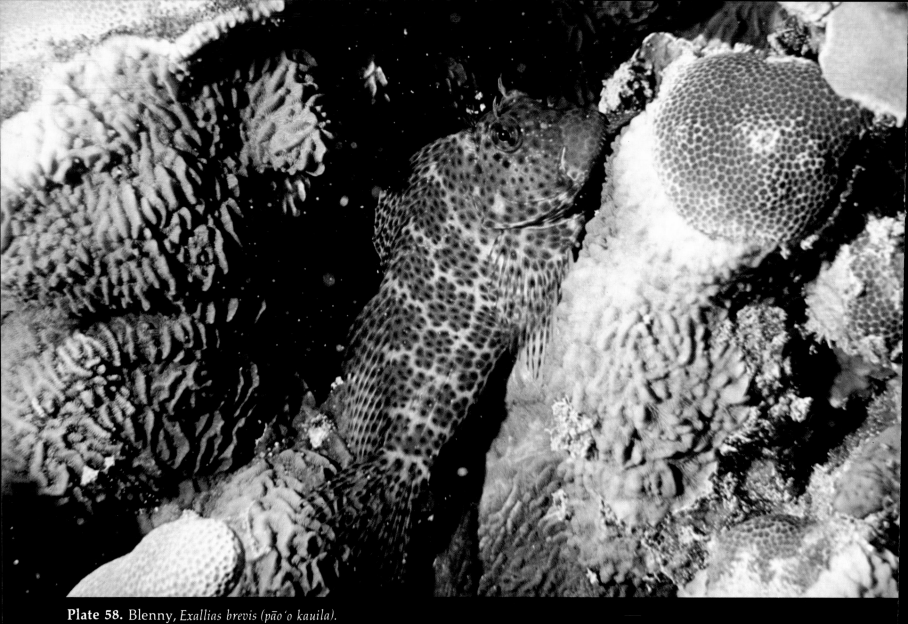

Plate 58. Blenny, *Exallias brevis (pāoʻo kauila)*.

Plate 59. Sabre-toothed blenny, *Runula ewaensis.*

pig). Others include *Melichthys niger*, called *humuhumu 'ele'ele* (dark *humuhumu*); *M. vidua*, called *humuhumu hi'u kole* (red-tailed *humuhumu*); and *Sufflamen bursa* (Plate 60), known as *humuhumu umauma lei* (*humuhumu* with leis on its chest).

The closely related filefishes are considered by some ichthyologists to comprise a separate family, the Monacanthidae. They can be distinguished from the triggerfishes by the position of the large first dorsal spine, which is farther forward—over the eyes in filefishes, as compared to behind the eyes in triggerfishes. Additionally, filefishes have a soft velvety body surface, in contrast to the hard, almost armor-like encasement of triggerfishes.

Early Islanders knew several kinds of filefishes as different forms of *'ō'ili* (to make an appearance). *Pervagor spilosoma* (Plate 61), called *'ō'ili 'uwī'uwī* (squealing *'ō'ili*), is a relatively common species that does not exceed a length of about 5 inches. Occasionally this fish appears in inshore areas in great numbers, with literally millions washing up onto the beach. In ancient days, this event was thought to foretell the death of a high chief (2). *'Ō'ili 'uwī'uwī* have never been valued as food, but because their flesh is very oily ancient Hawaiians regularly burned them as fuel.

Another common filefish is *Cantherines sandwichiensis*, called *'ō'ili lepa* (flag-bearing *'ō'ili*). This is a rather nondescript brown species with pink fins that grows to about 7 inches long.

Boxfishes

The boxfishes, family Ostraciontidae, are peculiar creatures. The body of each is encased in a solid bony box, so that swimming movements are confined to the fins and tail; understandably, they are not strong swimmers. Although seemingly vulnerable to predators, boxfishes show little concern when closely approached by a diver. Because at least some boxfishes secrete a poisonous substance from their skin (10), it may be that predators reject them as prey.

The most common boxfish in Hawaii is *Ostracion meleagris*, known as *pahu* (box). Most *pahu* are dark with white spots, but the large males are brilliantly hued with yellow and blue (Plate 62). The striking appearance of this fish, and the ease with which it is captured, would make the *pahu* a tempting target for those seeking aquarium specimens; however, this is a temptation to be resisted, because its poisonous secretions will kill other fishes with which it is confined.

88

Plate 60. Triggerfish, *Sufflamen bursa (humuhumu umauma lei).*

Plate 61. Filefish, *Pervagor spilosoma* (ʻōʻili ʻuwīʻuwī).

Pufferfishes

The pufferfishes include three distinct groups of species: the balloonfishes, which constitute the family Tetraodontidae; the sharpback puffers, which are considered by some to comprise a family distinct from the above, the Canthigasteridae; and the spiny puffers, or porcupinefishes, which comprise the family Diodontidae.

In old Hawaii, most puffers were known, collectively, as *'ōpū hue* (stomach like a gourd), or *kēkē* (pot-bellied). The name refers to their most outstanding characteristic: an ability to inflate themselves like a balloon by swallowing water or air. Another characteristic of these fishes presents a paradox: although the puffers are among the most poisonous of all marine animals, many of them are highly prized as food. Certain internal organs and the skin contain a potent nerve poison. Although the flesh is edible, a human who eats a piece that has been contaminated during preparation by poison from an adjacent organ has better than a 60 percent chance of violent death (31).

Arothron hispidus (Plate 63) is the most numerous of the balloonfishes in Hawaii; *Canthigaster jactator* (Plate 64) is the most common sharpback puffer; and *Diodon hystrix* (Plate 65) is the spiny puffer seen most often. Some balloonfishes grow to about 2 feet long, and some of the spiny puffers attain a comparable size, but the maximum length of sharpback puffers is only about 5 inches. All pufferfishes have sharp beaks and powerful jaws. These structures, which can seriously injure humans, are adaptations for feeding on hard-shelled organisms, like sea urchins and mollusks.

Anglerfishes

The anglerfishes, family Antennariidae, are unusual creatures. Those that live on the reef look more like a lump on the sea floor than a fish, an illusion heightened by their characteristic lack of movement. They rarely swim above the substrate; when one moves, the effort usually looks like an awkward crawl, with the animal using its jointed pectoral fins much like arms. But the most unusual feature of many anglerfishes is the modified first spine of their dorsal fins. This spine, which is far forward on the head, has transformed into a flexible "fishing pole," complete with a fleshy "lure," or bait, at the end. Waving this bait in front of its mouth, the anglerfish waits until a small fish is attracted. Once the little fish has approached within range, the anglerfish springs open its mouth and sucks the prey in.

Plate 65. Porcupinefish, *Diodon hystrix*.

Seven different species have been reported from Hawaiian reefs. Only a few are seen with any regularity, and none are abundant (10).

Perhaps the most common species is *Antennarius drombus* (Plate 66).

Plate 66. Anglerfish, *Antennarius drombus.*

The Invertebrates

Invertebrates are animals without backbones. They include the corals, jellyfishes, snails, marine worms, sea urchins, crabs, shrimps, lobsters, and many other forms. Because of the number and complexity of invertebrate groups on Hawaiian reefs, only a few that contain the larger, more common species are described in this book.

Jellyfishes, Sea Anemones, and Corals

Jellyfishes, sea anemones, and corals comprise a group of animals classified by scientists as the Cnidaria. The body form of all these animals is essentially a sac with a single opening surrounded by tentacles. In jellyfishes, called medusae, this saclike body form usually is bell-shaped and semitransparent; medusae usually swim freely in the water. On the other hand, the saclike bodies of both corals and anemones are formed as elongate cylinders, called polyps, that are attached to the substrate. The polyp of the sea anemone is usually large and solitary, whereas coral polyps often are small and clustered together in colonies. The tentacles of polyps and medusae carry stinging cells, some of which can painfully wound humans. Fortunately for Islanders, Hawaiian waters know few dangerous corals and sea anemones. The medusae should be avoided, however, because many of them can sting humans on contact.

The oldest islands of the Hawaiian chain, the Leeward Islands, are coral reef atolls built on volcanoes about 20 million years old. Sea and wind have long since eroded these volcanoes away, leaving the scattered bits of land and shallow water reefs we see today. With so much time to develop, some of the reefs here are Hawaii's richest. The newest island, Hawaii, contains the components of a coral reef community, but only as a relatively thin veneer over a base of volcanic rock. Reefs grow slower in the

Hawaiian Islands than elsewhere, perhaps because these northernmost of Pacific coral reefs are bathed in relatively cold water, and the main reef-building corals, species of *Acropora*, are not present.

Coral reefs are complex associations of vertebrates, invertebrates, and plants. Reef corals themselves exist in complex association with certain algae, which are a simple kind of plant. These algae live within the coral body-tissue and give to corals their hues of brown, yellow, pink, or blue.

An underwater swimmer in Hawaiian waters finds the most common corals to be species of *Porites, Pocillopora,* and *Montipora.* Collectively these are given the Hawaiian name *ko'a.* Islanders also gave this name to their fishing shrines, perhaps because these shrines frequently were constructed of coral skeletons. These skeletons are almost always white, in contrast to the various hues possessed by coral in life. Because of its visibility at night, *ko'a* rock was also important in marking trails.

Coral colonies consist of many polyps joined by thin layers of living tissue. Each polyp occupies a depression in the surface of the skeleton and lays down layer after layer of limy material, thus building a coral head larger and larger. New polyps appear between the old ones as the head grows. Coral colonies may form solid or branching heads, or they may encrust on surfaces. Species of *Porites,* which are yellow when living, occur in all three forms: massive (frontispiece), branching (Plate 3), and encrusting (Plate 60). Species of *Pocillopora*, however, grow only as branching heads that usually are brown, or pinkish tan in life (Plate 28). The branching form is the most popular for decorations, often being sold to tourists after receiving a coat of gaudy paint.

Not all corals build reefs or live in large colonies. One form, *Tubastrea aurea* (Plate 67), lives in small colonies in caves and rocky shadows. The brilliant hues of this species, perhaps the most colorful of Hawaiian corals, range from black to bright orange and red. Its polyps are fully expanded at night, but withdraw when exposed to daylight.

Mollusks

A wide variety of creatures comprise the group called Mollusca, or mollusks. Because the shells of mollusks have long been avidly collected the world over by laymen and professionals alike, the taxonomy of this group is better known than that of any other major group of animals, excepting the mammals and birds. The shell is the most outstanding characteristic of mollusks, even though it is reduced to a rudiment in some species, and is entirely lacking in a few others. Despite the prominence of the shell, the characteristics that unify members of this group are mostly in their soft parts. Among the most readily recognized are a muscular foot, generally used for locomotion, and folds of skin, called a mantle, which may cover the upper part of the animal's body.

Early Hawaiians used mollusks mostly for food, but also valued them for medicines, jewelry, horns, and fishing lures (1). In the Hawaiian language, shelled mollusks are known collectively as *pūpū*. Nowadays the word *pūpū* usually refers to an *hors d'oeuvre*, which probably relates to the long-standing use of such small mollusks as 'opihi (limpet) and pipipi (*Nerita*) as snacks. There are about 1,000 species of marine mollusks in Hawaii, ranging in size from the giant triton, *Charonia tritonis*, which may attain a length of 16 inches, to such tiny forms as *Tricolia variabilis*, a common beach shell that does not grow over one tenth of an inch long. Of the five major mollusk groups, three are not treated in this book: the clams, and other bivalved species (pelecypods); the chitons (amphineurans); and the tooth shells (scaphopods). For more information on these and other Hawaiian mollusks, see the work of the late Dr. Charles H. Edmondson, of the Bishop Museum (34), or the

Plate 67. Orange coral,
Tubastrea aurea.

field guide to shells by Percy A. Morris (35). In this book we consider the cephalopods, which include the squid and octopus; and the gastropods, which include the various snail-like mollusks so popular with collectors.

Cephalopods are curious creatures possessing a well-developed head that projects into a circle of eight or ten large, prehensile tentacles. Most do not have shells. Although only a few cephalopods are common in the Islands, these were well known to early Islanders.

The two major groups of cephalopods are the squids and the octopi. Two species of octopi, called *he'e*, are common on the reef: *Polypus marmoratus* and *P. ornatus* (Plate 68). Both are secretive by day and active at night. *P. ornatus*, known to Hawaiians as *he'e pūloa* (long-headed *he'e*), is known as "night squid" to torch fishermen, who find it abundant on the reef after dark (34). Islanders today commonly call the octopi "squid," even though the squid were recognized in ancient times as a distinct group of animals called *mūhe'e*. The squids, or *mūhe'e*, are not reef creatures, but free-swimming forms of the open ocean. Frequently they are taken when nets are placed to capture fishes swimming in the midwaters.

He'e and *mūhe'e* are still important as food in Hawaii, even though the flesh of both is tough and must be boiled or pounded before it is suitable for eating. Some early Islanders considered *he'e* and *mūhe'e* to be *'aumākua* (6) and in early times they served as remedies for certain illnesses. The word *he'e* means to dissolve, disperse, or put to flight, so probably this remedy was thought capable of driving away a malady (1). Several of the chants concerning these animals used in ancient times by *kāhuna* or fishermen are still heard today in Hawaiian songs and *hula*.

He'e are fished by several methods. In daylight, the simplest approach is to poke sticks into holes in the reef that harbor *he'e*, and then spear the animal when it is driven into the open (3). An ancient method was to dangle one or more spotted humpback cowrie shells on the end of a string over a hole known to contain a *he'e*. *He'e* have a particular attraction to this type of shell, and were captured when they left their holes and grabbed the cowrie (1, 3). Fishermen, using torches, hunt *he'e* on the reef at night. When the animal is encountered in exposed locations, it is speared or captured by hand. A frequently used method of quieting a writhing *he'e* is to bite it between the eyes. In handling these creatures, one should be careful to avoid their sharp parrotlike beaks, which can inflict painful injury. Early fishermen used the ink sac of the octopus (*'ala 'ala he'e*) as a basic ingredient in fish baits. These baits were prepared in several ways. Sometimes the ink sac was roasted, pounded, and then combined with red pepper, four or eight *ilima* flowers (odd numbers had no potency in Hawaiian lore), and salt (3).

Most mollusks in Hawaii belong to the group known as gastropods, which have a single shell (in contrast to the bivalved clam and its relatives). There are over forty families of gastropods in Hawaii, with shapes ranging from the long, slender shells of augers, to the cap-shaped shells of limpets. Some, the nudibranchs, have no visible shell at all. Gastropods live in a wide variety of places, from rocky sea cliffs above the water, dampened only by periodic surf spray, to deep water offshore. In between these extremes various members of the group live in sand, rocky crevices, under rubble, among *limu* (seaweed), and in most other submarine habitats.

A number of gastropods live on rocky shores at the water's edge. At low tide, when many are high and dry, they remain clamped tightly to the rocks. At high tide, when they are submerged or bathed by breaking waves, they move slowly over the rocks, scraping fine algae with a rasping type of tongue called a radula, a unique characteristic of gastropods. These animals are readily collected, and have long been sought after by Islanders. Among such mollusks are the tiny black nerite, family Neritidae, which Hawaiians called *pipipi;* and the littorines, family Littorinidae, called *pūpū kōlea.* These mollusks, which are less than an inch long, are usu-

ally boiled before being eaten. The limpets of the family Patellidae are larger animals. Several species in this group, called *'opihi,* are favorite foods, especially when eaten raw. One must move quickly to remove *'opihi* from the rocks, however, because the moment these animals are touched their foot secures a firm hold on the rock, and they then become very difficult to dislodge. *'Opihi* were considered by some early Hawaiians to be *'aumākua.* It was thought they were capable of calming heavy surf, so that fishermen could safely launch their boats, and return them to shore (1).

Many gastropods live in the sand, where they are inactive in daylight. Probably the largest of these is the helmet, *Cassis cornuta* (Plate 69), family Cassididae, called *'olē.* This large animal often grows to over a foot long, and not only is its flesh good eating, but its shell makes an excellent horn. By blowing into a small hole, filed at the end of the shell, one can produce a low resonant sound. Today, as in ancient times, the sounds produced with helmet and triton shells introduce many Hawaiian ceremonies. Other prominent sand-dwelling gastropods in Hawaii include the miter shells, family Mitridae, called *pūpū lei 'aha'aha;* the auger shells, family Terebridae, called *pūpū loloa;* and the stromb shells, family Strombidae, called *mamāiki.*

Members of several groups of gastropods remain hidden by day in rocky crevices, or under boulders. Included among these are the cowries,

family Cypraeidae, called *leho*. The tiger cowrie, *Cypraea tigris* (Plate 70), is the largest of our Hawaiian cowries; the 6-inch specimens that occur here are larger than any members of this species on other Pacific reefs (36). The highly polished, colorful shells of cowries make them especially attractive—their glossy finish is maintained by secretions from the animal's mantle. The mantle of the cowrie completely covers the animal's shell when it is active (Plate 70). Cowries lay their eggs in clusters, and the female spreads her foot over the eggs until they hatch. Mollusks hatch from their eggs as larvae that look very different from the familiar adult. After a time of drifting around in the midwaters as part of the plankton, the larvae settle to the substrate and transform into the forms we know.

Another denizen of coral or volcanic substrates is triton's trumpet, *Charonia tritonis,* family Cymatiidae, called *pū*. Its shell, inhabited by a hermit crab, appears in Plate 71. This large gastropod crawls about at night and searches for the sea stars (starfishes) and sea cucumbers that are its prey. *Pū* are not common in Hawaii, perhaps because they have long been sought intensively by collectors. Like the *'olē*, the *pū* is edible; and its shell was used by early Islanders as horns and drinking vessels.

Another prominent gastropod is the partridge tun, *Tonna perdix* (Plate 72), family Tonnidae, which grows to about 7 inches long. After spending the day secreted in the sand, the partridge tun emerges with darkness and hunts sea cucumbers on the reef. Although living specimens are not often seen, the shell is frequently found inhabited by the hermit crab *Dardanus.* When this lively crab encounters anemones of the genus *Calliactis,* it places them on the shell it inhabits. These anemones perhaps protect the crab by discharging stinging threads when disturbed. Unfortunately for the shell collector, the anemones leave scars on the tun shells, which diminish their value as specimens.

The cones, family Conidae, a group characterized by their conical shape, live in a range of different habitats, and offer the collector a variety of colorful shells. Cones are carnivores. Their sting is a nerve poison that the cone injects into its victims with a long, grooved implement derived from the radula. This sting can painfully wound humans. In fact, three local cones reportedly are potentially lethal to humans: *Conus marmoreus, C. textile,* and *C. striatus* (31). *C. marmoreus* and *C. textile* prey on other mollusks, and *C. striatus* preys on fishes.

Although it is the shell that draws the eyes of collectors to most mollusks, some of the most spectacularly beautiful animals in the sea belong to a group of gastropod mollusks that do not have visible shells at all. These are the nudibranchs or sea slugs, a large assemblage that includes many different families. Unfortunately, our illustration (Plate 73) depicts one of the less colorful nudibranchs inhabiting Island waters.

Plate 71. Shell of triton's trumpet, *Charonia tritonis* (pū), inhabited by the hermit crab *Dardanus punctulatus*.

Plate 72. Partridge tun, *Tonna perdix.*

Most are very small animals, one inch long or less, but a few grow to a length of one foot or so. Usually they occur on the sea floor, but some swimming forms glide through the mid-waters with graceful undulations. To be appreciated, nudibranchs must be seen in life; they do not preserve well, a fact that has made work on the taxonomy of the group difficult.

Shrimps, Crabs, and Lobsters

Animals comprising the group called crustaceans include the familiar shrimps, crabs, and lobsters. These creatures are characterized by being encased in a rigid, armorlike shell—an external skeleton. This affords the animal effective protection, but also creates problems in movement and growth. The armor is arranged in discrete sections, joined by articular membranes that permit movement between the different sections. The jointed legs of these animals are one of their outstanding features, and led early Hawaiians to call them "i'a that have feet like prongs" (1). Medieval knights might well have had crustacean anatomy in mind when designing their metallic armored suits, because they solved the problem of movement in the same way. Growth is attained by periodic shedding of the outer shell, a process called molting, during which the animal ruptures its body shell and then pulls itself out. The new shell, already formed under the old one, is still soft and pliable, and once free of its discarded encasement the animal quickly increases in size before its new shell hardens. Once this has happened, no further growth is possible until the next molt.

Even though the larger crustaceans are the ones sought after by fishermen, some of the very small forms are actually more important in the ecology of the reef. For example, the tiny copepods, which are only a fraction of an inch long, are the major food of many reef fishes. These minute crustaceans number many species on the reef and in the waters above, but are not considered in this book. Our concern here remains centered on those creatures that catch the eye of human visitors to the reef.

The shrimps, called by Hawaiians, collectively, 'ōpae kai, are prominent animals on the reef. One common species is the bandana prawn, *Stenopus hispidus* (Plate 74), family Stenopidae. During the day this animal hides in holes in the reef, but at night it emerges into the open and scavenges on fragments of dead animals. Sometimes this shrimp picks at the body of a living fish, and thus has been classified as a cleaner, along with the little wrasse, *Labroides phthirophagus*, discussed earlier. Bandana prawns usually occur in mated pairs. The female carries the developing eggs under her abdomen with specialized appendages. As was true of mol-

112

lusks, the young prawns hatch as larvae and drift in the midwaters for a time as part of the plankton, before settling to the sea floor and assuming adult form and habits.

Another common shrimp on the reef is the spiked prawn, *Saron marmoratus* (Plate 75), family Hypolytidae, called by Hawaiians 'ōpae kākala. This creature has many habits in common with the bandana prawn, although it is not known to clean fish. The male can be distinguished by its first pair of legs, banded red and grey, which are much longer than those of the female. The female is distinguished by the many tufts of hairs that she carries on her back and on her first pair of legs.

When swimming above the reef, one usually hears the crackling sounds made by snapping shrimps, family Alpheidae. These sounds, which the animals produce with a modified element of their single large pincer, perhaps function to frighten predators, or to communicate with others of their own kind. Snapping shrimps, too, occur in mated pairs.

The spiny lobsters, family Palinuridae, *Panulirus penicillatus* and *P. japonicus*, collectively called *ula*, do not have pincers; however, their bodies are heavily spined, and one who would hunt them is wise to use gloves. *Ula* scavenge for bits of dead animal flesh on the reef after dark; in daylight they retire to holes in the reef (Plate 76), where they often congregate in large numbers. During the breeding season, the fe-

male spiny lobster broods large clusters of pink eggs under her abdomen for about one month. Hawaiian law forbids the taking of these females at any time, and no spiny lobster can be taken during June, July, or August. They cannot be speared, and must exceed a weight of one pound to be legal game (37).

Another sought-after species, a member of the family Scyllaridae, is the slipper lobster, *Scyllarides squammosus* (Plate 77), called *ula pāpapa* (flat *ula*). This species grows to about a foot long, and is excellent eating. Two other slipper lobsters occur on Hawaiian reefs, but are too small to make hunting them worthwhile. Like the spiny lobsters, the slipper lobsters are nocturnal. During the day, they are even more secretive than *ula*, and are effectively hunted only after dark.

Perhaps the crustacean most highly prized as food in Hawaii is the Kona crab, *Ranina serrata*, family Raninidae, called *pāpa'i kua loa*. These animals are not abundant and are prized so highly that few manage to get into the markets. Kona crabs are denizens of sandy bottoms at depths exceeding 20 feet. During the day they are mostly buried in the sand, with only the front part of the body, and the eyes visible.

Although most crustaceans are nocturnal animals, with secretive habits in daylight, the shore crabs of the family Grapsidae typically crawl along the water's edge during the day. These are known by various Hawaiian names:

114

Plate 75. Spiked prawn, *Saron marmoratus* (ʻōpae kākala).

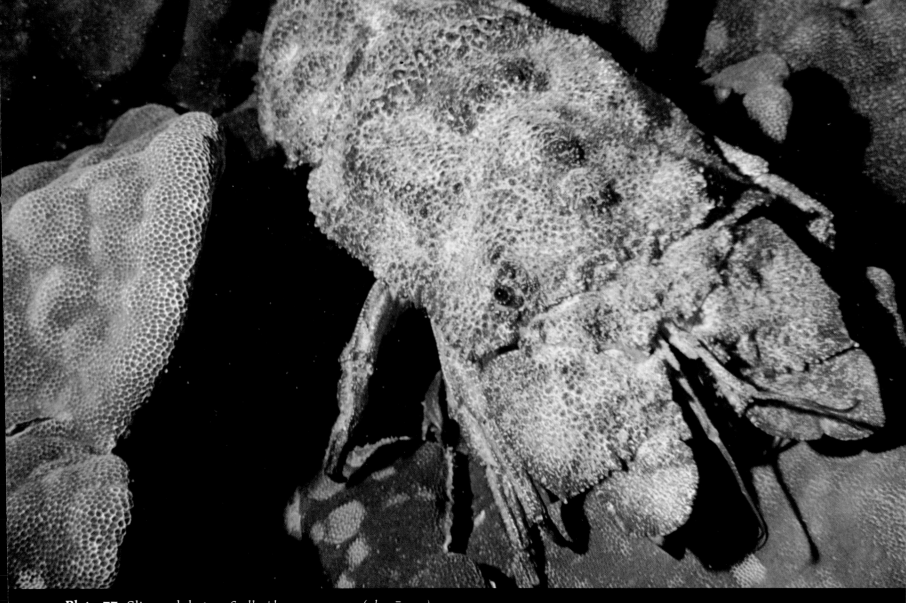

Plate 77. Slipper lobster, *Scyllarides squammosus (ula pāpapa).*

hīhī wai, *'alamihi*, and *'a'ama*. The *'a'ama* is *Grapsus grapsus*, a crab of the rocky shoreline that was used as a ceremonial sacrifice as well as for food in old Hawaii. Islanders caught this elusive crab with a line baited with *'opihi* (3). Several old chants concern the shore crabs, including:

Black crabs are climbing,
crabs from the great sea,
sea that is darkling.
Black crabs and grey crabs
scuttle o'er the reef. (8)

If one drags a large piece of coral rock out of the water, and breaks it apart on the beach, an amazing number of tiny crabs scurry out from among the debris. Most of these crabs belong to the family Xanthidae. Although most of the one hundred or so Island species of this family are small, one is the prominent 7-11 crab, *Carpilius maculatus* (Plate 78), which may attain a length of 6 inches. Xanthid crabs were eaten by early Hawaiians, with the exception of certain poisonous varieties called *kūmimi*, which were used by *kāhuna* in sorcery (18).

Another prominent group of crabs are the swimming crabs, of the family Portunidae. The last pair of legs on these crabs, modified as paddles, so effectively propel them through the water that they are recognized as the most powerful and agile swimmers of all crabs. Yet despite this natatory ability, many spend a large part of their time buried in the sand.

Burrowing in the sand is even more characteristic of the box crabs, family Calappidae. The body shells of crabs in the genus *Calappa* have wide winglike expansions that cover their legs. Several species, whose shells are covered with knobby protuberances, look like rocks until they move.

The ghost crabs, which comprise the family Ocypodidae, are prominent along the water's edge on sandy beaches. These crabs dig burrows in the sand from which they emerge and scavenge on bits of food among the debris that has washed up on the shore. Their burrows are readily recognized by the conical pile of sand next to each entrance. Ghost crabs are highly active animals that scurry across the sand with amazing agility. They owe their name to their bleached coloration, a feature that serves them well as camouflage against a background of white sand. Significantly, the ghost crabs living on the Islands' black-sand beaches are themselves dark grey.

Hermit crabs, *pāpa'i iwi pūpū*, comprise a specialized family, the Paguridae, that is adapted to life in abandoned gastropod shells. Whereas the abdomen of a typical crab is turned under the forward part of its body, that of the hermit crab projects backward as a soft, banana-shaped structure that fits into its adopted home. The front end of these crabs has the hard covering typical of other crustaceans, and when

Plate 78 7-11 crab, *Carpilius maculatus*

the crab has withdrawn inside its shell, its right pincer, larger than the left, serves to close the entrance.

As a hermit crab grows, it must periodically replace its borrowed shell with a larger one. When a prospective replacement is found, the crab carefully examines it with his claws. If the shell is satisfactory, the crab quickly transfers his vulnerable abdomen to the new home. Sometimes one crab will attempt to pull another from a particularly desirable shell.

Many hermit crabs inhabit shallow water, the most common of these being species of the genus *Calcinus*. These crabs often congregate under boulders, and when these boulders are overturned the crabs scurry in all directions. The largest of Hawaiian hermit crabs, *Dardanus punctulatus*, usually occupies the shell of the triton (Plate 71). This and other hermit crabs make good aquarium pets. Like most other crustaceans, hermit crabs readily accept bits of fish or mollusk meat, and a single piece, once a week, will keep one well and active.

Sea Urchins, Sea Stars, and Sea Cucumbers

Some of the most conspicuous invertebrates on Hawaiian reefs are members of the group known as the Echinodermata, a word that means spiny skin. Included here are the sea urchins, sea stars (starfishes), and sea cucumbers. Most members of this assemblage are encased in a skeleton of armorlike plates, on which are mounted spines or other projections. Yet the most striking characteristic of the group is the radial symmetry of their body plan; that is, their body parts are arranged symmetrically around a central point like spokes in a wheel. Thus they have a top and a bottom, but no front or back. However, each group of animals within the Echinodermata has developed distinctive variations on this generalized theme.

The sea urchins, or echinoids, are enclosed in globular skeletons, or tests, and most are heavily spined. The tests are hard, and contain many small holes through which protrude the muscles that move the spines, as well as the tube feet that give the animals mobility. Most sea urchins are herbivores that use a beaklike feeding apparatus to scrape fine algae off the surfaces of rocks and corals.

Several Hawaiian sea urchins have venomous spines. One of the more prominent of these is *Echinothrix diadema* (Plate 79), of the family Diadematidae, which the Islanders call *wana* (spiny). The spines of the *wana* are of two sizes: smaller, secondary spines are interspersed among larger, primary spines. The spines are

Plate 79. Sea urchin, *Echinothrix diadema (wana).*

brittle, and after puncturing the skin of an unfortunate fish or human they often break off deep within the wound. Thus, in addition to the immediate intense pain caused by the venom, the human victim is likely to suffer a secondary infection as well. The gonads of this species, like those of many other sea urchins, are sought after for food—a delicacy eaten raw. When collecting *wana*, fishermen place many of them together in a wire cage that is then shaken until all the spines have been broken off.

Another common large sea urchin, *Tripneustes gratilla*, has short spines that are not dangerous to humans. Called *hāwa'e* by Islanders, this species is a member of the family Toxopneustidae. *Hāwa'e* characteristically carry on their spines small stones, shells, or fragments of algae. Like those of the *wana*, the gonads of this species are sought for food.

One small sea urchin of the family Echinometridae is an important force in eroding reefs (38). This is *Echinometra mathaei* (Plate 80), called *'ina uli* (dark *'ina*). Color forms of this species range from pink to green to black. By continuously abrading a single spot on the reef, this urchin gradually hollows out a pocket that not only provides shelter to the urchin itself, but ultimately to many other organisms that find a haven there after the urchin's death. The extensive pitting of many reefs is the work of countless generations of *'ina uli*.

Another member of the family Echino-

metridae is the well-known slate-pencil urchin, *Heterocentrotus mammillatus* (Plate 81), called *'ina 'ula* (red *'ina*). The club-shaped spines of this urchin have long been used to make wind chimes. In addition, over the years, casual visitors to the reef, their attention drawn by this animal's unique characteristics, have thoughtlessly carried them ashore only to later discard them on the beach. The result of this unfortunate popularity is that in areas readily accessible to humans this creature is scarce.

A distinctive inhabitant of rocky shores, *Colobocentrotus atrata* (Plate 82), called *hā'uke'uke*, is another member of the family Echinometridae. This animal is highly specialized to a life clinging against rocks awash in the surf. The spines over the top of the flattened test are short and table-shaped; the longer spines around the margin of the test, along with the tube feet, secure a firm grasp on the rock. So tenacious is their hold that fishermen seeking them for food usually pry them loose with a knife in much the same way they collect *'opihi*. In ancient times the teeth of this urchin were valued as medicine (22).

In the popular mind, sea stars, or asteroids, are among the animals most characteristic of the sea. The familiar arms of sea stars are not appendages, as one might suppose, but are in fact extensions of the body; each carries a segment of the reproductive organs, digestive tract, and other body components. This fact

Plate 80. Sea urchin, *Echinometra mathaei* ('ina uli).

Plate 82. Sea urchin, *Colobocentrotus atrata* (hāʻukeʻuke).

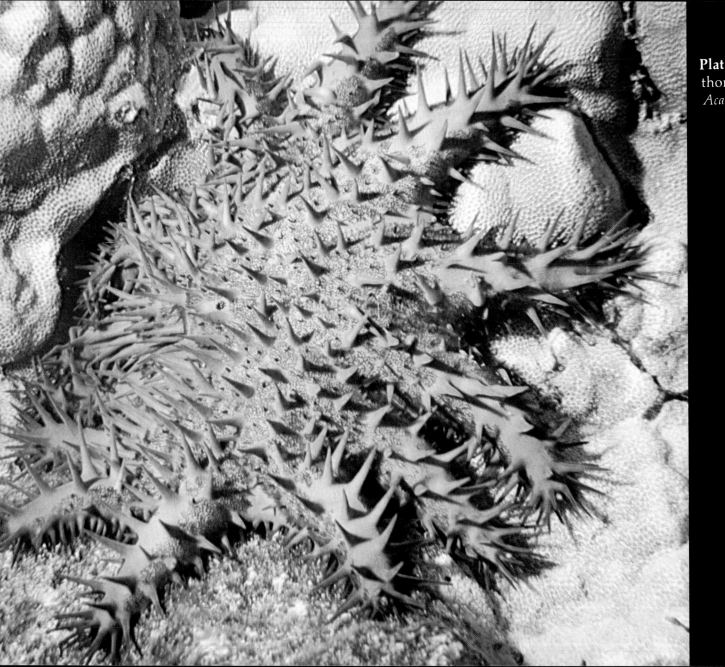

Plate 83. Crown-of-thorns, *Acanthaster planci.*

makes it easier to understand the remarkable ability of sea stars to regenerate parts of their bodies that have been lost. This ability is particularly well developed in species of *Linckia*, family Ophidiasteridae, a group with several representatives on Hawaiian reefs. These can be recognized by their smooth, leatherlike skin, and small central body with long, thin arms. Even a piece of the arm, when torn from one of these creatures, can reorganize itself into a whole new animal. Thus, if one is fragmented, a number of new animals will result.

Although several species of sea stars are common residents of the reef, none have Hawaiian names so far as is known. Perhaps this is because none are suitable for human food.

A sea star that has attracted considerable notoriety in recent years is the crown-of-thorns, *Acanthaster planci* (Plate 83), family Acanthasteridae. Within the last decade this animal has appeared in exceptionally large numbers at certain places in the Pacific Ocean; and beause it preys on coral there has been concern about the long-lasting effect the increased predation might have on coral reefs. However, there is no evidence that this animal poses any unusual threat to Hawaiian reefs. The crown-of-thorns should not be handled because it has venomous spines.

The arms of the pin-cushion star *Culcita novaeguineae* (Plate 84), family Oreasteridae, are fused together. This sea star is another predator of corals, but one that supplements its diet with fine mats of algae. Often it carries within its body a parasitic fish of the family Carapidae. This knifelike fish lives in the sea star's gonads, entering through the mouth of its host.

The sea cucumbers, or holothurians, are elongated, often sausage-shaped creatures that lie with little movement on the sea floor. When their body plan, as an echinoderm, is compared with that of the sea urchins, or sea stars, sea cucumbers can be thought of as lying on their sides. The sea urchin and sea star both have their mouths on their undersides, facing the substrate, and their anuses are directly opposite, on the top of their bodies. In contrast, the mouths and anuses of sea cucumbers are located at either end of their long drawn-out bodies. With one side thus consistently lying against the substrate, the external body features of sea cucumbers have secondarily varied from the radial symmetry of the generalized echinoderm to a superficial bilateral symmetry. Usually on that side of the body that rests on the sea floor the tube feet are well developed, whereas on surfaces not in contact with the substrate these structures are reduced to wartlike protuberances, or are lost completely. The sea cucumbers further differ from other echinoderms in that their skeleton is reduced to minute plates embedded in their integument, and as a feeding mechanism they have a circle of tentacles around their mouth. These tentacles in many

Plate 84. Pin-cushion star, *Culcita novaeguineae.*

species shovel a continuous load of sand or mud into the mouth, and as this material moves through the gut the animal digests from it a great assortment of nutritive matter.

Members of the family Holothuridae are called *loli*. One representative, *Holothuria atra* (Plate 85), is a common sea cucumber on Hawaiian reefs. Its black leathery integument usually is covered by a fine layer of sand. Hawaiian legend claims that the *loli* originated from the dead body of a worm, called 'Enuhe, an evil *'aumakua* that lived in a cave at Ka'u, on the island of Hawaii.

When *loli* are disturbed, like many other sea cucumbers, they disgorge their viscera. The organs are emitted as a tangled mass of sticky material, probably as a deterrent to predators. After evisceration, the lost organs are soon regenerated. In some parts of the Pacific, natives deliberately agitate large sea cucumbers, and then collect the disgorged organs for food. The bodies too of many sea cucumbers are highly regarded as food, usually after being boiled and then dried or smoked. In the early years of this century, all species of *loli* were an important food in Hawaii (3), but today *Stichopus tropicalis*, family Stichopodidae, is favored.

Sea cucumbers of the genus *Opheodesoma*, family Synaptidae, are especially elongated, and have body walls that are very thin and flexible. These snakelike creatures feed mostly on decaying matter that clings to seaweeds. One species, *O. godeffroyi* (Plate 86), is a denizen of exposed rocky coasts, where it is seen mostly at night. Another, *O. spectabilis*, is a pinkish form occurring mostly in protected areas, like Kāne'ohe Bay.

Collecting Invertebrates

Many invertebrate species are sought after as trophies by both residents and visitors in the Islands. This is because so many of them have colorful or unusual skeletons that are readily preserved. The shells of mollusks and the skeletons of stony corals are particular favorites. Despite the relative ease with which these trophies can be preserved, countless specimens, upon spoiling, have been discarded by people who were unfamiliar with these procedures. For such individuals, we offer some suggestions.

Mollusk shells can be placed in a plastic container with pine oil, or some other detergent, for about two weeks. The detergent will decompose the meat in the shell, or at least loosen it sufficiently so that it is easy to remove. Visitors to the Islands will find that such a container, properly sealed, is readily carried home in lug-

Plate 85. Sea cucumber *Holothuria atra (loli).*

Plate 86. Sea
cucumber,
Opheodesoma godeffroyi.

gage. After removing the shell from the detergent, it should be washed in cold running water. If odor persists, the shell should be returned to the detergent. After washing, all water should be wiped from the shell, as water sometimes corrodes the finished surface. Shells should never be boiled.

Corals should be washed in freshwater, then placed in a 10 percent bleach solution. After soaking for a week or so, the coral should be washed thoroughly, preferably by water under pressure, and then dried in the sun until the smell of bleach is gone. Corals or shells should not be washed with vinegar because this substance dissolves away the lime skeletons and shells.

Most invertebrates are nocturnal animals that occur under cover during the day. Therefore, in daylight a collector is most successful if he looks under rocks, or stirs through the sand. However, if one employs these techniques it is vitally important that all overturned rocks or other structures be replaced in their original positions. Many animals make their homes in these rocks, shielded from sun and predators, and many of them will perish unnecessarily if the rocks are left overturned.

Bibliography

1. Malo, D. 1903. Hawaiian antiquities. Bernice P. Bishop Museum, Special Publication 2. 366 p.

2. Titcomb, M., with M. K. Pukui, collaborator. 1952. Native use of fish in Hawaii. Memoir 29 of the Polynesian Society. 162 p.

3. Cobb, J. N. 1902. The commercial fisheries of the Hawaiian Islands. U. S. Commission of Fish and Fisheries, Commissioner's Report for 1900-1901: 383-499.

4. McAllister, J. G. 1933. Archaeology of Oahu. Bernice P. Bishop Museum, Bulletin 104. 201 p.

5. Beckwith, M. W. 1917. Hawaiian shark aumakua. American Anthropologist 19: 503-517.

6. Emerson, J. S. 1892. The lesser Hawaiian gods. Hawaiian Historical Society, paper 2. 24 p.

7. Richards, D. K. 1941. Men beat the sea and its gods. Honolulu Star-Bulletin, 1 March 1941.

8. Emerson, N. B. 1909. Unwritten literature of Hawaii. Bureau of American Ethnology, Bulletin 38. Smithsonian Institution. 288 p.

9. Beckley, E. M. 1887. Hawaiian fisheries and methods of fishing with an account of the fishing implements used by natives of the Hawaiian Islands. U. S. Fish Commission, Bulletin 6: 245-256.

10. Gosline, W. A., and V. E. Brock. 1960 Handbook of Hawaiian fishes. Honolulu: University of Hawaii Press. 372 p.

11. Manu, M. 1901. Ku'ula, the fish god of Hawaii. Translated by M. K. Nakuina. Hawaiian Almanac and Annual for 1901: 114-124.

12. Kamakau, S. M. 1870. Story of Hawaii. *Ke Au Okoa.* 6 January 1870.

13. Hobson, E. S. 1968. Predatory behavior of some shore fishes in the Gulf of California. U. S. Fish and Wildlife Service, Research Report 73. 92 p.

14. Eibl-Eibesfeldt, I. 1965. Land of a thousand atolls. London: MacGibbon and Key. 194 p.

15. Hobson, E. S. 1963. Notes on piloting behavior in young yellow jacks. Underwater Naturalist 1 (4): 10-13.

16. Helfrich, P., and A. H. Banner. 1960. Hallucinatory mullet poisoning. Journal of Tropical Medicine and Hygiene 63: 86-89.

17. Anonymous. 1912. Noted places on Lanai. *Ka Nupepa Kuokoa.* 31 May 1912.

18. Pukui, M. K., and S. H. Elbert. 1971. Hawaiian dictionary. Honolulu: University of Hawaii Press.

19. Gosline, W. A. 1965. Thoughts on systematic work in outlying areas. Systematic Zoology 14: 59-61.

20. Randall, J. E. 1961. Two new butterflyfishes (family Chaetodontidae) of the Indo-Pacific genus *Forcipiger*. Copeia 1961 (1): 53-62.

21. Wheeler, A. 1964. Rediscovery of the type specimen of *Forcipiger longirostris* (Broussonet) (Perciformes-Chaetodontidae). Copeia 1964 (1): 165-169.

22. Beckwith, M. W. 1932. Kepelino's traditions of Hawaii. Bernice P. Bishop Museum, Bulletin 95. 206 p.

23. Reinboth, R. 1962. Morphologische und funktionelle Zweigeschlechtlichkeit bei marinen Teleostiern (Serranidae, Sparidae, Centracanthidae, Labridae). Zoologische Jahrbücher Abteilung für Allgemeine Zoologie und Physiologie der Tiere 69: 405-480.

24. Losey, G. S., Jr. 1971. Communication between fishes in cleaning symbiosis. *In* Aspects of the biology of symbiosis, ed. T. C. Cheng, p. 45-76. Baltimore: University Park Press.

25. Bardach, J. E. 1961. Transport of calcareous fragments by reef fishes. Science 133 (3446): 98-99.

26. Winn, H. E., and J. E. Bardach. 1959. Differential food selection by moray eels and a possible role of the mucous envelope of parrotfishes in reduction of predation. Ecology 40 (2): 296-298.

27. Reinboth, R. 1968. Protogynie bei Papageifischen (Scaridae). Zeitschrift für Naturforschung 23b (6): 852-855.

28. Fornander, A. 1918-1919. Fornander collection of Hawaiian antiquities and folklore. Bernice P. Bishop Museum, Memoirs, vol. 5. 721 p.

29. Rice, W. H. 1923. Hawaiian legends. Bernice P. Bishop Museum, Bulletin 3: 137 p.

30. Jones, R. S. 1968. Ecological relationships in Hawaiian and Johnston Island Acanthuridae (surgeonfishes). Micronesica 4: 309-361.

31. Halstead, B. 1959. Dangerous marine animals. Baltimore: Cornell Maritime Press. 146 p.

32. Strasburg, D. S. 1960. The blennies, p. 277. *In* W. A. Gosline and V. E. Brock, Handbook of Hawaiian fishes, Honolulu: University of Hawaii Press.

33. Hobson, E. S. 1969. Possible advantages to the blenny *Runnula azalea* in aggregating with the wrasse *Thalassoma lucasanum* in the tropical eastern Pacific. Copeia 1969 (1): 191-193.

34. Edmondson, C. H. 1946. Reef and shore fauna

of Hawaii. Bernice P. Bishop Museum, Special Publication 22, 381 p.

35. Morris, P. A. 1966. A field guide to shells of the Pacific coast and Hawaii. Boston: Houghton Mifflin Co. 297 p.

36. Kay, E. A. 1961. On *Cypraea tigris schilderiana* Cate. Veliger 4 (1): 36-40.

37. State of Hawaii. 1960. Revised laws of Hawaii. Division of Fish and Game Regulation Number 22.

38. Hodgkin, E. P. 1960. Patterns of life on rocky shores. Journal of the Royal Society of Western Australia 43: 35-43.